The Origins of Satellite Communications, 1945–1965

Smithsonian History of Aviation and Spaceflight Series

Dominick A. Pisano and Allan A. Needell, Series Editors

Since the Wright brothers' first flight, air and space technologies have been central in creating the modern world. Aviation and spaceflight have transformed our lives—our conceptions of time and distance, our daily routines, and the conduct of exploration, business, and war. The Smithsonian History of Aviation and Spaceflight Series publishes substantive works that further our understanding of these transformations in their social, cultural, political, and military contexts.

THE ORIGINS OF SATELLITE COMMUNICATIONS 1945–1965

DAVID J. WHALEN

SMITHSONIAN INSTITUTION PRESS
Washington and London

The poem “Syncom” is reproduced with permission of Punch Ltd.

Copy editor: D. Teddy Diggs
Production editor: Robert A. Poarch

Library of Congress Cataloging-in-Publication Data
Whalen, David Joseph.
The origins of satellite communications, 1945-1965 / David Joseph Whalen.
p. cm. — (Smithsonian history of aviation and spaceflight series)
Includes bibliographical references and index.
ISBN 9781935623601
1. TK5104.W48 2002 2. Artificial satellites in telecommunications—History. I. Title. II. Series.
621.382'5—dc21 2001057613

British Library Cataloguing-in-Publication Data is available

Manufactured in the United States of America
09 08 07 06 05 04 03 02 5 4 3 2 1

Contents

Tables and Figures

Tables

Figures

Preface

Challenging the conventional wisdom—the assumption that government (politics rather than economics) developed satellite communications because industry could not—is the purpose of this book. That part of the story told here is of the invention of two kinds of communications satellites by AT&T and Hughes Aircraft Company, followed by government intervention to control that technology for political purposes. The primary rationale for government intervention appears to have been a desire to enhance foreign and domestic prestige in the cold war era by demonstrating a practical application of space technology. In addition, a combination of antitrust, antibusiness, and anti-AT&T sentiment contributed to the final outcome.

The postwar period was characterized by a dramatic increase in government funding of research and development (R&D) and a dramatic improvement in the economy. These events resulted in an assumption that a specific government investment in commercial R&D could have a specific, *cost-effective* economic result. This assumption ignored the market completely. Although the "push" of a new technology can occasionally develop new demand, it is at least as common for the "pull" of the market to encourage entrepreneurs to develop new technology.

The conventional wisdom suggests that government communications satellite R&D produced the satellite communications industry. This is not true. Most of the technologies were developed before the 1957 launch of *Sputnik 1*. Perhaps more important, the market for satellite communications—transoceanic communications—already existed and was growing at a rapid rate. The economic value of communications satellites was recognized by private companies, whose response was to begin investing in communications satellite R&D.

A major factor affecting the development of satellite communications was AT&T's dominance of domestic and international telephony in the United States. Many companies were interested in building communications satellites. But most of these companies were thwarted by monopsony (one buyer) rather than monopoly (one seller). AT&T, as the dominant U.S. telephone company—certainly the only one heavily involved in transoceanic telephony—was the single potential buyer of communications satellites. AT&T had shown a tendency over the years to buy only from its internal supplier, Western Electric. None of the other suppliers had much of a chance to sell their ideas, no matter how good, to AT&T.

Hughes had a better idea: a lightweight geosynchronous satellite. Hughes even tried to form partnerships with GTE and ITT to perform an end run around AT&T. Unfortunately, the partnerships never materialized. AT&T was uninterested in a Hughes communications satellite. For a while, no one else seemed interested either. Government intervention eventually made it possible for Hughes Aircraft Company's better idea to triumph in the marketplace, but was this an intended effect or an accidental effect? Would the Hughes technology have eventually triumphed anyway? The government did not "invent" communications satellite technology—did it even provide a level playing field? Was refusing launch service to AT&T fair? Was minimizing AT&T's participation in Comsat fair?—especially after AT&T's investment of tens of millions of dollars in satellite communications and similar investments in the development of basic technologies such as transistors, solar cells, and traveling-wave tubes? The process by which the government intervened in the development of satellite communications seems unnecessary (industry was funding its own R&D) and unfair (AT&T was excluded after having funded the entire Telstar program).

Five outcomes of government intervention will be traced in the pages that follow: (1) the forcing of AT&T out of satellite communications; (2) the establishment of Comsat/Intelsat; (3) the positioning of Hughes as the dominant commercial satellite manufacturer; (4) the establishment of geosynchronous earth orbit (GEO) as the preferred orbit; and (5) the dominance of U.S. manufacturers of satellites and launch vehicles.

Though not described at length in this work, government intervention in satellite communications development must be seen as part of the cold war. This was not "policy for technology"—that is, policy whose purpose is to advance technology. Government intervention in satellite communications is an example of "technology for policy"—that is, the use of technology to advance public policy. In this case, the U.S. government wanted to show the world that it was ahead of the Soviet Union in space and show the American taxpayers that their money spent on the space race would have practical applications.

Finally, this work does not describe all of the events that led to the emergence of satellite communications. This is just one of many stories, but it is a story that has not been told. Many individuals—"space cadets," defense intellectuals, cold war politicians, engineers, businesspeople—and many institutions helped make satellite communications and the global village a reality. This story is about the contributions of engineers, businessmen, and market-driven industry.

Acknowledgments

Any historical work will always incur debts to the many who assisted with the research, commented on the research and conclusions, and/or edited the text. This work is no exception. The initial research was done at the University of Maryland Libraries, the George Washington University Library, and the NASA libraries at the Goddard Space Flight Center and NASA Headquarters. Primary source material was obtained from the NASA History Office Archives, the National Archives, the AT&T Archives, and the Hughes Aircraft Company Archives (Office of Record Management). I thank the staff of all of these libraries and archives for their assistance. Additional assistance was obtained from Roger Launius, NASA Historian, and Lee Saegesser, NASA Archivist. Thanks go also to Andrew J. Butrica, for editing assistance and for organization of a NASA conference on satellite communications in 1995; that conference allowed me the opportunity to present my findings to an informed audience. Special thanks must go to Helen Marie Gavaghan, who shared her research on Hughes and AT&T, and to the Sloan Foundation, which funded her research. At George Washington University, John M. Logsdon, Nicholas Vonortas, and William H. Becker were always encouraging. Faith Keenan provided editorial advice at critical periods. Many of the participants in this chronicle—John R. Pierce, John H. Rubel, Sid Metzger, John Townsend, Simon Bennet, Richard Marsten, Sam Goldfarb, Richard White, and many others—were generous with their time and helped to make this a better work. Many friends provided advice and encouragement from time to time. My wife, Mary Anne Whalen, put up with boxes and stacks of paper strewn over the house, with strict instructions not to touch anything lest it become disorganized.

Any errors are mine, but there would have been many more without the kindness of friends and colleagues. I thank them all.

Communications Satellite Chronology

October 1945	Arthur C. Clarke publishes the article "Extra-Terrestrial Relays" in *Wireless World.*
May 12, 1946	RAND publishes the study "Preliminary Design of an Experimental World-Circling Spaceship."
March 1952	John R. Pierce publishes the article "Don't Write; Telegraph" in *Astounding Science Fiction.*
1954	As a science-fiction author, John R. Pierce is invited to give a talk on space to the Princeton section of the Institute of Radio Engineers (IRE). He chooses to talk about satellite communications—a subject that was "in the air" at the time.
1956	First transatlantic telephone cable laid (TAT-1) by AT&T and the British Post Office.
October 4, 1957	Launch of *Sputnik 1.*
March 3–4, 1959	Congress holds two days of hearings on "Satellites for World Communication."
March 1959	Pierce and Kompfner satellite communications article in *Proceedings of the IRE.*
August 1959	Leroy C. Tillotson, of AT&T, designs an MEO satellite.
December 1959	The Department of Defense Notus program is reoriented. All resources are dedicated to the twenty-four-hour satellite program.
1959	Harold A. Rosen, Donald D. Williams, and Tom Hudspeth,

	of Hughes Aircraft Company, design a GEO communications satellite.
July 11, 1960	AT&T outlines its plans for a global communications system.
August 12, 1960	*Echo I* is launched into a 1,000-mile circular orbit.
August 18, 1960	The first Courier satellite fails, due to a Thor first-stage malfunction.
October 4, 1960	*Courier Ib* is launched successfully (it fails after eighteen days in orbit).
October 12, 1960	T. Keith Glennan delivers a speech in which he states that private industry should develop satellite communications.
October 21, 1960	AT&T files with the FCC for permission to launch and operate a communications satellite.
December 8, 1960	Abe Silverstein, NASA director of space flight programs, objects strongly to the presence of private companies in the communications satellite business.
January 4, 1961	Request for Proposal (RFP) for a NASA communications satellite (Relay) is released.
January 19, 1961	The FCC grants AT&T authorization to launch an experimental communications satellite.
March 1961	The FCC opens Docket 14024 soliciting views on the "administrative and regulatory problems" associated with a commercial satellite communications system.
May 8–10, 1961	The House Committee on Science and Astronautics holds hearings on communications satellites.
May 18, 1961	RCA is awarded the contract to build Relay.
May 24, 1961	The FCC favors a communications satellite system jointly owned by international carriers.
May 25, 1961	President John F. Kennedy makes a speech challenging the nation to commit to a manned moon landing within the decade. Included was a commitment to global satellite communications.
July 27, 1961	NASA and AT&T enter into agreements for the reimbursable launch of Telstar.
August 11, 1961	NASA signs a sole-source contract with Hughes to build Syncom.
January 11, 1962	The Kerr bill (S2650), favoring ownership of the global satellite communications system by the communications carriers, is introduced into the Senate.

January 27, 1962	The administration bill (S2814), favoring broad-based ownership, is introduced into the Senate.
February 26, 1962	The Kefauver bill (S2890), favoring government ownership, is introduced into the Senate.
May 22, 1962	The Director of Defense Research and Engineering study (by Ralph Clark) recommends cancellation of the Advent program.
July 10, 1962	AT&T successfully launches *Telstar 1.*
August 17, 1962	The amended "Kerr" House bill passes the Senate by a vote of 66 to 11.
August 31, 1962	President Kennedy signs the Communications Satellite Act of 1962.
December 13, 1962	*Relay 1* is launched by Thor-Delta.
December 1962	The Conference of European Postal and Telecommunications Administrations (CEPT) forms a committee to study the issue of joining a U.S.-led global communications system.
February 1, 1963	The Communications Satellite (Comsat) Corporation is incorporated in the District of Columbia.
February 14, 1963	*Syncom 1* is launched (it fails after the AKM firing).
May 7, 1963	*Telstar 2* is launched successfully.
July 26, 1963	*Syncom 2* is launched successfully.
December 1963	Hughes Aircraft Company proposes a commercial version of Syncom, which could be launched in 1965.
January 21, 1964	*Relay 2* is launched successfully.
April 1964	The London meeting of the European Conference on Satellite Communications proposes that COMSAT be the manager of the Intelsat consortium.
June 2, 1964	Comsat public shares are offered and immediately snapped up.
August 19, 1964	The Intelsat agreement is opened for signature.
August 19, 1964	*Syncom 3* is launched on a Thrust-Augmented-Delta (TAD).
April 6, 1965	*Early Bird* is launched, on schedule.
May 13, 1965	ABC files with the FCC for permission to launch a television relay satellite.

1. Introduction

The Billion-Dollar Technology

Following initial development by NASA[, c]ommunications satellites were quickly spun off to the private sector. . . . U.S. government investment in communications satellite research in the earliest days of the space program not only allowed development of increasingly sophisticated satellites for the United States, but for the rest of the free world. —Marcia Smith (Congressional Research Service, Library of Congress), "Civilian Space Applications," 1989

Most histories ascribe [the] pioneering role to NASA, which spent hundreds of millions of dollars on communications satellites during the 1960s and early 1970s. . . . But the enabling technologies for communications satellites, with the exception of rocketry, were developed by private industry before serious government interest in communications satellites. —Peter Cunniffe (MIT), "Misreading History," 1991

In late 1964, after the successful launches of the Telstar, Relay, and Syncom experimental satellites and before the launch of the Comsat-Intelsat *Early Bird*, James E. Webb, administrator of the National Aeronautics and Space Administration (NASA), asked his staff: "How did we get so much communications satellite technology for so little money?"[1] His question has never been adequately answered, in spite of several unpublished, NASA-sponsored histories written in the 1960s and independent attempts later. Because of the evident success of the NASA communications satellite program of the early 1960s, the conventional wisdom has portrayed this event, or series of events, as a technology-transfer success story in which NASA developed the communications satellite technology that was efficiently "spun off" to the private sector through the

Communications Satellite Act of 1962. This interpretation, summarized above by Marcia Smith,[2] is challenged by other analysts—such as Peter Cunniffe, also quoted above[3]—who view NASA's role instead as an intervention in an ongoing process.

The answer to Webb's question and a revision of the conventional wisdom might be found in the multibillion-dollar business of telecommunications. Funding has always been available for new telecommunications technology. The advantages of new technology have (almost) always been recognized. AT&T's comparative advantage through most of its history has been based on the control of patents—patents the company bought in the late nineteenth and early twentieth centuries and developed in-house later—especially after 1925. As early as 1954, John R. Pierce of AT&T was able to compare communications satellites with submarine telephone cables and ask: "Would a channel 30 times as wide [as TAT-1] be worth . . . a billion dollars?"[4] Compared with the TAT-1 (Trans-Atlantic Telephone #1) submarine cable technology (36 telephone channels for $36 million) then under construction, a 1000-telephone-channel satellite would have been worth $1 billion. In 1959 AT&T, with net sales of over $7 billion, was in a better position to fund communications satellite research and development (R&D) than NASA, whose entire budget that year, its first full year of operation, was only a few hundred million dollars. In 1962, during the Communications Satellite Act hearings, many witnesses and members of Congress asked how billions of dollars of government R&D could be allowed to benefit private industry. Those claims of billions of dollars spent were never justified—such sums had never been spent by the government on satellite communications.

At the close of the twentieth century, expenditures on satellite communications were larger than the entire NASA budget. Most of the world's population today has access to satellite television. The "global village" has become a reality. NASA apologists cite this as an example of the returns on investments in NASA. But is the federal government truly responsible for this new industry, or are market forces responsible? Is the conventional wisdom correct, or is this an example of the myth of continuous progress through government funding of high-technology R&D fostered by the New Deal, World War II, and the cold war? What are the real sources of the technological and economic development of satellite communications?

The point of view presented here is that satellite communications was always commercially viable. In contrast to the conventional wisdom, this book will argue the following:

1. Industry initially developed communications satellite technology with private funds.

2. Industry had sufficient private funding in 1957–61 and great market demand for international communications services.
3. The federal government intervened in communications satellite development in 1961 in order to (a) take political credit for this new technology, (b) control development, and (c) prevent an AT&T monopoly.

To set the stage for an analysis of the development of satellite communications, the following paragraphs will examine the interrelationships of market economics, government, and technology development.

Market Economics and Technology Development

Technology development is one of the major factors in economic growth. This was particularly obvious in the industrial revolution as completely new industries came to dominate the economy. Two of these, textiles and transportation, radically changed the U.S. economy. Samuel Slater's cotton mill (1790) and Eli Whitney's cotton gin (1793) allowed the United States to participate in the expanding high-technology textile industry at the end of the eighteenth century; later, Mississippi steamboats in the West (1816), canals in the East (e.g., Erie Canal, completed 1825), and eventually railroads everywhere began to knit together the huge U.S. domestic market in the early nineteenth century.

By the Civil War, the United States may have been the second-largest manufacturing economy in the world. In the period after the Civil War, with the country united coast to coast by the railroad and telegraph, new inventions were constantly entering the market. Thomas Alva Edison became a symbol of the enterprising, technically competent American: inventing new devices and profiting from them. The pantheon of nineteenth-century inventors also includes Samuel F. B. Morse, Elisha Gray, and Alexander Graham Bell. Edison and Gray differed from Morse and Bell in that they were managers of invention factories. Although Edison liked to maintain the fiction of the lone inventor, he and Gray were running large establishments whose goal was invention. By the first quarter of the twentieth century, large corporations—such as General Electric, AT&T, and Dupont—were operating their own invention factories. At least in the United States, technology development was seen as a source of profit by lone inventors, entrepreneurs, and major corporations.

Economists,[5] historians,[6] and technologists have studied technology development for many years. Technologists and some historians have concentrated on invention: the origination of a new device or process. Economists have seemed to focus on innovation: the introduction of a new device or process to the market. In the days of Edison, the two processes were closely related—the inventor and the innovator were the same person. By the early twentieth century, however, it was clear that the processes were separate.

Frederic M. Scherer, a Schumpeterian economist, wrote one of the classic articles on invention and innovation.[7] In a 1965 historical essay on the development of the steam engine, Scherer suggested that James Watt was the inventor of the (separate condenser) steam engine but that Matthew Boulton was the innovator who brought it to market. Scherer concluded, "Innovation [may be] more sensitive to economic variables than invention is."[8] Or as *The Economist* put it thirty years later: "As a general rule, a firm will not bother to innovate unless it thinks it can steal a march on the competition and, for a while at least, earn higher profits."[9]

In the twentieth century, economists debated whether innovation was endogenous (market pull) or exogenous (technology push). There was some agreement that incremental innovation was dominated by market pull but that revolutionary innovation was dominated by technology push. What large, innovative companies wanted to know was: "How do we foster innovation?" The late eighteenth century saw American industrialists stealing English technology to build American textile mills and other new devices and processes. But there was also native market pull and inventive push, as seen in Whitney's cotton gin. In many U.S. industries, foreign inventions were improved and made to fit local circumstances. America's incremental improvements were seen, especially by foreign observers, as due to more widespread education in America. By the early twentieth century, the R&D departments of major corporations had established strong links to universities such as MIT. Industry was convinced—before World War II—that R&D and education were keys to economic success.

Government and Technology Development

From the beginning of the Republic, when Thomas Jefferson and Alexander Hamilton debated their differing agricultural and industrial views of the American future, there have been disagreements and concerns over the extent of government support of industry.[10] The classic rationale for government involvement in science and technology has been as a supporter of basic research that has no immediate economically appropriable market value (i.e., low returns to the researcher) but that has large returns to society at large over the long term. In addition, certain functions that are inherently governmental, typically defense-related, are seen as appropriate for the government to support from invention through innovation and diffusion. Less clearly, there are areas of market failure in which the supply side of a market cannot support the necessary research—often because the market is too fragmented for any one actor to support R&D. The classic example has been agriculture. Individual farmers could not do the appropriate research, but the land-grant colleges could.

The first elements of industrial policy in the United States were high tariffs and the patent system. The high tariffs generated so much revenue that a campaign to fund internal improvements (roads, bridges, canals, telegraphs, and finally railroads) began in the early nineteenth century. By the mid-nineteenth century, most elements of science and technology policy seemed to be aimed at settling the frontier rather than, or perhaps in addition to, promoting manufactures. From the time of the Lewis and Clark Expedition until late in the nineteenth century, explorations and surveys provided knowledge of the potential mineral and agricultural wealth of the nation—as well as mapped the routes to that wealth. The basic research aspect of R&D during much of the nineteenth century consisted of low-level government (federal and state) and philanthropic support of the sciences, and the activities of individual inventors. In addition, two federally funded science and engineering educational institutions were founded in the nineteenth century: the U.S. Military Academy at West Point (civil and military engineering) and the U.S. Naval Academy at Annapolis (mechanical engineering).

World War I saw the beginnings of stronger government interest in R&D (the National Advisory Committee on Aeronautics—NACA—dates from this period), but the advent of peace was followed by a return to industry-dominated commercial R&D, supplemented by government laboratories. World War II proved, or seemed to prove, that massive government intervention could force technological development. The salient characteristic of the post–World War II U.S. innovation system was the mobilization of R&D in the service of national priorities. These priorities themselves were products of the cold war rivalry with the Soviet Union. During most of this period, the United States dominated the global R&D arena. The two major areas of government R&D funding were (1) electrical machinery and (2) aircraft and missiles.

The emphasis on space and missile R&D in support of national priorities is suggestive of one aspect of the development of satellite communications: government intervention seems to have been motivated primarily by defense, foreign policy, and prestige goals, not by an interest in technology development or an interest in the economy. The government contribution to the development of satellite communications was, at most, a spin-off from national priorities.[11] The defense buildup of the 1950s and 1960s probably did lead to U.S. dominance of the electronics and aerospace industries—although it is debatable whether that dominance was caused by direct support of R&D or by massive procurement contracts.[12]

Communications Satellite Technology

Three major technologies were (and are) required for satellite communications: launch vehicles (rockets), ground stations, and satellites. All satellite launch vehicles have their roots in the military missile programs of the postwar period. Several different launch vehicles were available in 1960, but all had limitations. By 1960 the NASA Thor-Delta, combining the Air Force Thor IRBM (intermediate range ballistic missile) and modified Vanguard upper stages, was on its way to becoming the most reliable satellite launch vehicle. It is ironic that the Vanguard, which failed spectacularly to launch the first U.S. satellite in 1957, should prove to have had more effect on later generations of launch vehicles that the more successful Redstone-Jupiter, which launched *Explorer 1* in 1958. All three stages of the Vanguard—the Viking first stage, the Aerobee second stage, and the solid third stage—had peaceful sounding-rocket origins. Unfortunately, the initial payload capability of the Thor-Delta was quite small—about one hundred kilograms to LEO (low Earth orbit) (200–1,000+ kilometers). The 1960 Atlas-Agena could put about 1,000 kilograms in LEO. The Atlas-Centaur was expected to be able to place about four thousand kilograms in LEO—but unfortunately, development problems made the Atlas-Centaur unreliable until the late 1960s.

Ground stations were widely available at the beginning of the space age. Descended from radar, communications, and radio astronomy activities, the antennas, receivers, and transmitters needed for satellite communications were existing components—but were often quite expensive. In the early days of satellite communications, individual ground stations were often more expensive than satellites. Of particular importance for the ground stations were low-noise receivers. The satellite signal, though extremely weak, was relatively noise-free. Most noise was added by the ground-station receiver. MASER (microwave amplification by stimulated emission of radiation) amplifiers and parametric amplifiers (both very low noise receivers), often cryogenically cooled, became available at just the right time.

Launch vehicle development was a military priority, and ground-station technology was well understood, but satellite technology was still in development. Since at least 1954–55 the specific technological questions to be answered before a communications satellite system could be built had been elaborated and debated.[13] The first of these questions was whether the satellites should be *passive* mirrors that simply reflected radio signals or *active* microwave repeaters that received the transmitted signal from the ground, amplified it, and retransmitted the signal to a distant ground station. It seemed clear that active satellites were better but that passive satellites might be much cheaper and simpler.

The second, and more complex, question concerned the satellites' orbit. Satellites in low orbits would simply whiz by in a few minutes. Continuous global communications might require hundreds of satellites. The ground stations would need complex tracking systems capable of following a satellite moving from horizon to horizon in minutes. Multiple antennas would be required at every site. The satellites might be simple, but the ground system would be complicated and more expensive.

Satellites in orbit 36,000 kilometers above the surface of Earth would have twenty-four-hour periods. They would revolve about Earth in the same time that Earth took to rotate—thus appearing to be stationary (more or less) in the sky, simplifying ground-station design. Three such twenty-four-hour or GEO (geosynchronous Earth orbit) satellites could cover all of the globe except for the polar regions.[14] There were several problems with the GEO system, however. First, the distance was so great—a slant range of 40,000 kilometers rather than the 3,000 kilometers of LEO systems that transmitter power would have to be almost two hundred times as great. Alternatively, high-gain satellite antennas and sophisticated satellite pointing systems (attitude-control systems) would be required. Second, maintaining the satellite in one geographic location would require the orbit to be continually modified to correct for perturbations caused by gravitational effects and solar radiation pressure. Third, it took more energy to get to GEO. A launch vehicle that could put 400 kilograms in LEO could put only 100 kilograms in GEO. GEO satellites made for simple, inexpensive ground systems but complicated satellites. The MEO (medium Earth orbit), about 10,000 kilometers above the surface of Earth, seemed a reasonable alternative. In MEO, satellites would still rise and set on a regular basis, but movement would be somewhat statelier than that of the "whizzing" LEO satellites. The slant range to the satellite would normally be about 10,000 kilometers, but like GEO satellites and unlike LEO satellites, MEO satellites would spend much of their visible time at relatively high elevation angles; thus, dense atmospheric columns would not be traversed.

In contrast to government support of rocketry, almost all of the basic inventions relating to communications satellites (and communications generally) originated years earlier in industry, primarily at AT&T's Bell Telephone Laboratories (BTL, or Bell Labs). Bell Labs, established in 1925, was perhaps the strongest of the industrial R&D laboratories.[15] In addition to close links with research universities, Bell Labs had a strong scientific component to its research. Studies of the properties of semiconductors led to the discovery of the transistor (1947) and the solar cell (1953), as well as improved diode rectifiers. Vacuum tube technology was well established at BTL and included most of the development process of the traveling-wave tube (TWT). Electronics and communications sciences were the main areas of research, but many other areas of science and engineering were studied at Bell Labs. As a government contractor, BTL had worked on missile technology. As the nation's leading communications company, it led research in microwave communications, TWTs, transistors, solar cells, and other technologies necessary to satellite communications. Perhaps more important, AT&T had a need for a high-capacity transoceanic communications link. The company could profit from satellite communications.

The transistor, the MASER, the TWT (a very wideband, very linear amplifier), the solar cell, and many other technologies were usually well known to engineers at the other electronics companies such as RCA, General Electric, ITT, Philco, and Hughes—often because they hired engineers who had worked at Bell Labs. The single most important technological development unique to communications satellites was the orbit- and attitude-control scheme developed by D. D. Williams under the direction of Harold Rosen at Hughes Aircraft Company in 1958–60. This scheme made simple, lightweight, geosynchronous satellites possible. None of these technologies were invented under government contracts.

Individuals, Institutions, and Controversies

Although it can be argued that satellite communications appeared when it did because of a wide variety of socio-economic-technological trends, several individuals and institutions were instrumental in shaping its actual development and implementation. Within industry, AT&T and General Electric were large, powerful organizations capable of influencing governments and markets. Somewhat smaller organizations, such as RCA and Hughes, made major contributions to satellite communications but had much less power to manipulate governments and markets. Within the government, the most powerful force was Congress. The White House had its traditional "bully pulpit," but government regulatory power resided in the Federal Communications Commission (FCC)—often reminded that it was an instrument of Congress. NASA, at the beginning of the period in question (1958–65), was a small organization with no real influence until the decision to go to the Moon.

One of the complicating factors in the analysis of the origins of communications satellite technology is the market power of AT&T in the 1950s and 1960s. AT&T did not have a complete monopoly of telephone services, but its size allowed it to dominate the economics, the policy, and the standards of the telephone industry. Many in the government, both in Congress and in the White House, were uncomfortable with AT&T's monopoly power. While the government worried about monopoly, high-technology communications equipment manufacturers worried about monopsony. AT&T was the sole customer for telephone technology—technology that it preferred to develop at Bell Labs and manufacture at Western Electric.

Equally important—perhaps more important—than the contributions of institutions were the contributions of individuals. Arthur C. Clarke was the "godfather" of satellite communications; John R. Pierce of AT&T was one of the "fathers" of satellite communications, and Harold A. Rosen of Hughes was the other "father."[16] Many others made contributions that shaped the course of satellite communications development. The first two NASA administrators, T. Keith Glennan and James Webb, are obvious contributors. President John F. Kennedy and Vice-President Lyndon Johnson made space important to U.S. politics, creating the conditions under which satellite communications developed after January 20, 1961. A host of others also made contributions, including John Rubel (Department of Defense), Sidney Metzger (RCA and Comsat), Leonard Jaffe (NASA), and Siegfried Reiger (RAND and Comsat).

Several controversies erupted among those involved in the technology development. The first of these controversies was between AT&T and the federal government. AT&T felt that it had pioneered Earth-station technology, satellite technology, and international coordination—using its own funds—only to see the fruits of its efforts taken away by legislation.

The Williams orbit- and attitude-control patent created a second controversy, this one between NASA and Hughes. NASA attempted to claim rights to the Hughes-Williams patents on the grounds that the Williams invention was "reduced to practice" using government funding. Hughes had actually built a prototype satellite using its own funds. After decades, the legal issues were finally resolved in favor of Hughes.[17]

A third controversy erupted between NASA and Congress. Once the Communications Satellite (Comsat) Corporation was formed, Congress saw no need to continue using public funds to develop technology for a private profit-making organization. This resulted in the reorientation of the Advanced Syncom satellite program in the mid-1960s to include noncommunications applications—hence the new name: Applications Technology Satellite (ATS). Just before the launch of *ATS-6*, Congress finally canceled all remaining NASA communications satellite technology programs. NASA and certain elements of industry and Congress fought to resurrect these programs, finally achieving success with the Advanced Communications Technology Satellite (ACTS).

Yet a fourth controversy surrounds claims by Comsat and Intelsat (International Telecommunications Satellite Organization) to an unlimited monopoly on the provision of satellite communications services—both civilian and military.

This monopoly claim was first raised in 1963 when the Department of Defense (DoD) began its procurement of what eventually became IDCSP (Interim Defense Communications Satellite Program). After much argument, DoD won its case that military needs were unique. Comsat, on the other hand, received a commitment that "nonmilitary" defense communications would be carried by commercial carriers. Comsat also attempted to prevent the establishment of domestic and regional communications satellite systems. The Nixon administration opened the way for domestic systems in 1970 with its "open skies" policy. The Reagan administration, with its "separate systems" policy, eventually opened the industry to all comers.

Finally, there was controversy within NASA. The memos criticizing the NASA-sponsored "Thompson history" of communications satellite development (discussed below) and other earlier memos are revealing of the divisions within NASA. Some individuals argued that technology development was most important. Others argued that not enough attention was being paid to policy development. Some argued that satellite communications development was something that only NASA could implement. Others argued that industry had already begun this development in 1960. NASA Administrator Webb himself seems to have been looking for a history of technology development that emphasized not only the importance of government support and control but also the relatively low cost of technology development with this support and control.

A common theme among most of these controversies is the relative importance of government contributions to the development of satellite communications. The government was not attempting to fund basic satellite communications research—that had already been done by industry. Military communications satellites were handled by DoD. The government's attempt to take over civilian communications satellite development had none of the conventional motivations. Neither basic research nor inherently government functions were at the heart of this intervention. Although satellite communications could not be construed as an inherently governmental function, senior government officials were always lamenting the absence of a communications satellite policy. To some extent, there was a market failure associated with the monopoly (and monopsony) power of AT&T, but this does not seem to have been the primary motivation of the government. A "policy for technology" would have attempted to develop and demonstrate all technological options (especially GEO)—as the ATS program did in the late 1960s and early 1970s. The government sought to control the development of satellite communications technology and the institutions that would provide a global satellite communications system for political purposes. This was "technology for policy." Without government intervention, an AT&T-dominated MEO satellite consortium would almost certainly have provided this service. Greater prestige in cold war foreign policy was obtained by making global satellite communications a U.S. government project rather than a commercial project by the largest capitalist monopoly: AT&T.

Historiography: Developing the Conventional Wisdom

At least three different impulses have motivated studies of communications satellite technology development. The first was James E. Webb's desire to know how so much technology had been developed in so little time for so little money. The second impulse was the desire to justify NASA's push to reenter the communications satellite R&D field with ACTS. Finally, there is the impulse behind this study: a desire to show that industry's contributions to communications satellite technology development have not been adequately appreciated.

Eugene Emme, NASA chief historian, began a chronology of satellite communications in 1963.[18] In 1964 Edgar W. Morse began a NASA-sponsored history of communications satellites but quickly realized that this task was too large and decided to concentrate on the Syncom program.[19] This work resulted in a rather well-crafted article in which Morse pointed out that much of the Syncom effort took place before NASA became involved. He also covered the maneuvering that took place in the spring and summer of 1961 to get the Hughes Syncom satellite project approved. Apparently the urgings of Len Jaffe, NASA's communications director, were not enough; backing from John Rubel, director of defense research and engineering, was required. The result was a joint NASA-DoD program that resulted in three launches, two successes, and a system for use by DoD in Southeast Asia. Morse clearly saw, and reported, the importance of private funding and DoD support.

In 1965 George Thompson wrote a more ambitious NASA-sponsored history: "NASA's Role in the Development of Communications Satellite Technology."[20] This work was an analysis of DoD and NASA activities and a brief analysis of the economics of satellite communications. The main focus was NASA satellite R&D, especially Syncom R&D. What is surprising is the negative reaction of senior NASA management to Thompson's history. Webb wrote two memos complaining about the study.[21] One of Webb's complaints was that he wanted a history of communications satellite technology development, not a history of "who did what." Thompson allocated many pages and words to discussing AT&T's investment and the difficulty of keeping AT&T out of satellite communications and keeping NASA in charge. Meanwhile Arnold Frutkin, NASA assistant administrator for international affairs, claimed credit for saving money by getting foreign countries to build their own Earth stations, with no mention of AT&T.[22] Walter Radius, in policy planning, reiterated Webb's complaint that technology development was not really covered.[23] These comments were seconded by John D. Iams, also in policy planning,[24] and Lou Vogel, NASA executive officer.[25]

It is not clear that anyone liked Thompson's work, but Leonard Jaffe and Robert G. Nunn, two of the NASA men closest to the action, had the mildest complaints. Jaffe suggested that a sharper focus on technology (rather than policy) and a "more concise, narrative style" would have improved the work.[26] Nunn had only general comments, although he did suggest that policy was downplayed.[27] Many of those who commented on the work suggested that the interplay of NASA, DoD, and foreign partners should be emphasized. Except for Webb's comments on AT&T, there is no mention of private industry. John R. Pierce of AT&T was sent a draft of the study. Pierce quoted Napoleon: "History is a fable agreed upon." Pierce pointed out AT&T's contributions but was clearly convinced that NASA was out to "take credit" for everything AT&T had done.[28] A somewhat later (1968) study[29] examined the origins of the ATS program in the Advanced Syncom satellite program, but no examination of technology development seems to have been completed, and none of the early histories were published.

Two policy-oriented studies were published in the 1970s. In *The Politics and Technology of Satellite Communications,* Jonathan F. Galloway was interested in examining the reaction of decision-makers and policy-makers to technological innovation, not the development of new technology through government initiative.[30] Galloway's study is almost exclusively about policy. As one of the first book-length analyses of communications satellites, Galloway's study shows little interest in technology or economics. The conventional wisdom was already at work, suggesting that the Communications Satellite Act of 1962 had somehow "created" the industry. What was interesting about communications satellites was not the development of unique technology but rather the development of the policies that controlled the technology.

A few years after Galloway's work, Delbert D. Smith's *Communication via Satellite: A Vision in Retrospect* was published.[31] Those years had seen the success of Comsat-Intelsat, the cancellation of NASA's communications satellite R&D, and the beginnings of domestic satellite communications. Smith's study was sponsored by NASA's Office of Applications, headed by Leonard Jaffe, formerly responsible for communications research at NASA Headquarters. In the foreword, Jaffe presented several conclusions drawn from the events:

> (1) The capability of government, and in particular of NASA, to dramatically advance technology is past debate.

(2) That public benefit has been and can be a result of a government research and development approach has also been established.
(3) What has not been identified in the historical sequence is the meaning of the developmental process for the role of government—both present and future—in regard to new technologies and the rise of technologies already developed.[32]

Jaffe seemed to be primarily concerned that the noneconomic benefits of applications satellites had been rejected along with the canceled ATS program. He bemoaned the emphasis on relevance, costs, and benefits. He also stated, "The federal role in technology development and applications must expand if technological innovation is to flourish."[33]

In his conclusion, Smith lamented the 1973 NASA decision to withdraw from communications satellite experimentation. He stated that (as of 1976) not all potential satellite applications had been implemented. Smith noted, "There is little evidence that the private sector is willing or able to advance research and development along the lines dictated by public interest requirements."[34] Yet satellite communications was ubiquitous, and most of the new technology was the product of private industry—what was Smith implying? The answer seems to be that nonprofit applications, such as those pioneered by *ATS-6,* were not developing quickly. Smith's work is his own, but it is probably not unreasonable to assume that his conclusions were heavily influenced by Jaffe and that this was the beginning of the lobbying effort to start a new program—a program that eventually became ACTS. An article written by Ellis Rubinstein, "Dollars vs. Satellites," appeared in the October 1976 *IEEE Spectrum* magazine.[35] The primary point made here was that the private sector would not (initially, could not) make investments in R&D at the required rate. The second point was that foreign governments, especially the Japanese and Europeans, would make these investments and take away the U.S. economic advantage. The point most often made by Jaffe, in this magazine interview and others at the time, was that many communications satellite applications were not good commercial investments—there was no foreseeable profit—and that only government would fund R&D directed toward these noneconomic applications of communications satellites.

The article raised the hackles of John R. Pierce, the retired Bell Labs director of communications research, who had encouraged the Echo and Telstar programs. In a letter to Rubinstein, Pierce complained that too much credit had been given to NASA.[36] Pierce pointed out that the Echo program had been recommended by Pierce (AT&T), Rudolf Kompfner (AT&T), and William H. Pickering (head of the Jet Propulsion Laboratory) as an added experiment to a

NACA Langley Research Center atmospheric experiment using a large balloon. AT&T's internally funded Telstar program was the first true communications satellite. In a similar letter, another AT&T official pointed out that most current communications technology had come from Bell Labs.[37] The response from the *Spectrum* article author, Rubinstein, is a symptom of the pervasiveness of the conventional wisdom. He claimed that he had spoken to "more than 25 people—in NASA, in private industry, and in academia." His manuscript had been reviewed at NASA—"right to the top."[38] His assumption is clear: NASA would know all the facts. The AT&T engineers continued the correspondence among themselves, lamenting that AT&T's advances were not recognized. Pierce's final letter to Rubinstein had the same sorts of comments that his mid-1960s letter to Eugene Emme had contained: he was convinced that the history was already carved in stone—and it was *wrong*![39]

In the mid-to-late 1980s, two short articles on space policy and history contained expressions of the conventional wisdom. The first of these was written by Pamela Mack, a historian of science and technology whose later (1990) book, *Viewing the Earth*,[40] portrayed a program (Landsat) that NASA was pushing on users. (Perhaps this was an example of "technology push"? Satellite communications, on the other hand, was a user program—"market pull"—preempted by NASA.) In her 1981 article, she made the important point that in 1960–61, "NASA insisted that a government-funded and government-controlled experimental communications satellite . . . be planned first."[41] NASA wanted to *control* satellite communications development—not just contribute to it. According to Mack, "NASA proceeded with research on a geosynchronous communications satellite too advanced for the private companies to risk on their own."[42]

The second article was written by Marcia S. Smith, a space policy specialist with the Congressional Research Service (CRS) of the Library of Congress. In her 1989 article, the two pages devoted to communications satellites provide an almost complete exposition of the conventional wisdom:

> (1) Following initial development by NASA[, c]ommunications satellites were quickly spun off to the private sector,
> (2) although NASA remained active in research and development until 1973. A six year period followed in which the private sector was expected to take responsibility for developing new technologies, and when it did not, NASA reentered the research and development area in 1979 [ACTS]. [A corollary to both of the above is the assumption that industry could not afford this R&D.]
> (3) U.S. government investment in communications satellite research in the earliest days of the space program not only allowed development of in-

creasingly sophisticated satellites for the United States, but for the rest of the free world.

(4) Through the creation of INTELSAT, world-wide communications became a reality,

(5) and U.S. satellite manufacturing companies profited.

(6) Today, the United States continues to be the world leader in manufacturing communications satellites, but European industry has become very strong in this area and Japan will probably pose a competitive threat in the future.[43]

In 1991 the Brookings Institution published a book on federal R&D support of commercial technologies. One of the case studies in the book described the NASA ATS program, which was considered one of the few successes.[44] While most of the study centered on ATS, with only a few pages on the early systems, the authors repeated many elements of the conventional wisdom. The most important of these was the assumption that industry would not develop new communications satellite technology because (1) industry would not be able to appropriate the benefits, (2) such development was too risky, and (3) industry simply did not have the money. The second element of the conventional wisdom presented in this case study was that existing advances came from government R&D. The authors attributed the Indian, Arab, Brazilian, Mexican, and Indonesian satellite systems to the ATS program, especially *ATS-6*.[45]

Another 1991 contribution to the satellite communications history-and-policy literature was the MIT report "Misreading History" by Peter Cunniffe. His primary purpose was to "identify what role the federal government has played and should play in performing or funding research and development for the commercial communications satellite industry."[46] Cunniffe began by outlining the origins of commercial communications satellites and coming to conclusions in conflict with the conventional wisdom. These conclusions are summarized as follows:

(1) [M]ost of the technology for commercial communications satellites has been developed by private firms.

(2) [G]eneric space technologies that were developed with NASA and Defense Department funding have contributed to the success of commercial communications satellites, often through technology sharing between government and commercial projects. Government funding for development of and improvement in launch vehicles was by far the most important contribution.

(3) [T]he military contribution has mostly been the scale economies and generic technology sharing that has provided U.S. satellite manufacturers

with a competitive advantage over foreign rivals, and sometimes over one another.

(4) [G]overnment procurement may have been more important in providing early help to the new international comsat system than R&D sponsorship.[47]

Policy History

In the period between the two world wars, José Ortega y Gasset described the coming "rebellion of the masses" as a new social phenomenon that could best be understood through history.[48] More recently, in the mid-1980s, Ernest R. May, of the Harvard School of Government, and his colleague Richard E. Neustadt described their feelings about Washington decision-making: "We sensed around us . . . a host of people who did not know any history. . . . Yet we also saw that despite themselves Washington decision-makers actually used history in their decisions, at least for advocacy or for comfort, whether they knew any [history] or not."[49] Technology policy has a problem common to most policy-making: the policy-makers know very little history. Richard R. Nelson, a prominent economist who has examined the processes of technological development and attempted to advise the government on proper technology policy, observed in 1982: "[We] were keenly struck by a sense of *de novo* about the [1979 industrial innovation] review. It seemed to us that the discussion did not adequately recognize that the United States had a long history of policies aimed at stimulating innovation in various economic sectors."[50]

Perhaps the principal reason for the persistence of an incomplete view of the origins of communications satellite technology has been the lack of an in-depth history of the market. So far, the focus has been on political contributions to that development. Satellite communications may be the most significant product of the space age, given its global impact. It is therefore important that a history of the development of this technology be written.

The view that the government developed communications satellite technology and, after demonstrating the technology, transferred this technology to industry reflects the philosophy that the government's function in the development of technology is to provide the initial impetus (generic technology), after which industry can take over. The success of the communications satellite technology transfer has been used to justify other government-sponsored technology-transfer activities. Often when the transfer is attempted, as with the Landsat remote-sensing technology, there is great surprise when it does not work. NASA was surprised by its success in the development of communications satellite technology. In both the Landsat failure and the satellite communications success, the government seems to have ignored the effect of the market. The market for satellite communications existed in 1961. The market for remote sensing did not exist in 1984—more accurately, it did not exist at the prices that the NASA Landsat system required to cover its very high costs. This study will attempt to show the business and technological story behind satellite communications. The political story will intrude, but the emphasis will be on industry and technology.

If the major problem with the conventional wisdom is that it ignores the industrial origins of communications satellite technology, we would be equally wrong to ignore the results of government intervention. A different path would have been taken if the government had not intervened. Both technologies and institutions would have been different. There will be lessons to be learned from a close examination of the historical development of the technology. Many of the major players are still living—and in some cases are still disputing the manner in which the events discussed here played themselves out.

This study is primarily a narrative history of the development of the technology, an analysis of that history, and an examination of the consequences of the actions of the government and industry from 1945 to 1965 and in the post-1965 evolution of the industry. Chapter 2 covers the period 1945–57, from the end of World War II to the launch of *Sputnik 1*. It examines the technological roots of satellite communications technology in the postwar period. Chapter 3 covers the period 1957–60, from the launch of *Sputnik 1* to the last days of the Eisenhower administration. It examines technology developments and also the public policy events of 1960. Chapter 4 covers the period 1961–62, from the NASA Relay RFP, through the contracts for Telstar, Relay, and Syncom, to the passage of the Communications Satellite Act of 1962. It examines government intervention in satellite communications: first, with the NASA decision to take control of the launch manifest; second with the NASA decision to enter the field of satellite development itself by funding the NASA-RCA Relay and taking over the funding of further development of the NASA-Hughes Syncom; and third, the decision by Congress to overrule the FCC and create its own "chosen instrument" to develop the global communications satellite system (Comsat). Chapter 5 covers the period 1962–63: the building and launching of Telstar, Relay, and Syncom. Chapter 6 covers the period 1963–65, from the incorporation of Comsat to the December 1965 decision that Comsat's "basic system" would be geosynchronous (*Intelsat 4*). Chapter 7 reaches some tentative conclusions and briefly examines the later evolution of satellite communications.

Industry contributed to economic development by direct funding of invention and innovation. Government—at least the U.S. government—contributed to technology development by providing a fertile environment: patent policy, education, and protective tariffs. Could the government replace industry in the direct funding of invention and innovation—providing a faster route to economic development? This idea developed in the United States after World War II in spite of a national mythology emphasizing individual initiative and the lone inventor. In relation to this study we must ask: did the government invent communications satellites, or did John Pierce and Harold Rosen?

On May 25, 1961, President Kennedy made an important foreign policy speech that called for new rockets, a Moon landing (*Apollo 11*), and a global communications satellite system. These promises were kept. The space race is generally seen to have been another area of superpower rivalry during the cold war. The events of April 1961—the Bay of Pigs and the orbital flight by the Soviet Yuri Gagarin—are difficult to separate from the May 1961 speech promising to land a man on the Moon and develop a global satellite communications system. National prestige and the prestige of the Kennedy administration are generally seen as the major issues at stake. However, NASA and Congress saw another issue. There was a need to justify the space race from a more pragmatic point of view. The science content of the space race had been disparaged by the scientific establishment, especially by academic scientists, but there was still a chance to highlight the technological advances, particularly those technological advances that seemed to show economic payoffs. Communications satellite technology could be used to justify the billions of dollars that would be spent on the space program in the future.

Parts of this story have been told in the past—usually from the point of view that the government was responsible for the development of satellite communications. My intent here is to provide the historical evidence—emphasizing industry and technology—that will restore a more realistic view of this development. Clearly, some sort of industrial policy is required to foster economic and technological development, but there must be a recognition that industry is closer to this development process than government can be. Profit and loss provide a natural feedback path. Companies doing the right things get richer. Companies doing the wrong things get poorer. Given a chance to create a competitive advantage for themselves, companies will spend the R&D funds needed to bring advanced products to market. Aerospace companies will always be willing to accept government contracts to develop new technology, but none will be willing to risk their own funds unless they are convinced that they will profit from their R&D investments. There will always be market externalities that require government intervention. In the case of satellite communications, this externality was the dominant power of AT&T. There were other mechanisms for curing this problem; they need not have included the exclusion of AT&T from this industry and certainly should not have resulted in the government taking credit for the technological achievements of private industry.

2. From World War II to Sputnik

Inventing Communications Satellites

There has been a great deal said about a 3,000-mile high-angle rocket. In my opinion such a thing is impossible. —Vannevar Bush, December 1945, quoted by R. Cargill Hall in Eugene Emme's *The History of Rocket Technology*

Perhaps the most useful and obvious possibility associated with the space station is the world-wide television coverage envisaged by Clarke. So far such a system has been thought of in connection with large manned space stations, but it is the opinion of the writer that an automatic station can give such a service to humanity, at least temporarily, until the manned space stations can be constructed. —Eric Burgess, "The Establishment and Use of Artificial Satellites," *Aeronautics*, September 1949

World War II produced both the technologies and the technology policy that led to communications satellites and communications satellite policy. This new technology policy carried with it assumptions that were critical to later attempts to understand the genesis of satellite communications. One of the tools used to defeat fascism was American science and technology. Victory was not the result of superior U.S. military skills—skills that many Europeans considered mediocre at best. It was the joining of scientists, engineers, technicians, and industrial managers to build the radars, airplanes, and atomic bombs that won the war. The secret weapons of German technology had succumbed to U.S. technological and industrial know-how. What would later be called the "military-industrial complex" was created during World War II. Perhaps more important, success in war made the American public believe that this partnership between industry and government could provide the answers to other problems—especially economic problems. In the words of Frederick Lewis Allen, who chronicled the prewar period from a business-economic point of view: "When World War II came along, we discovered that if Washington jammed the accelerator to the floor boards the engine began to run smoothly and fast. And when the war was over, and Washington released the accelerator, it still hummed."[1]

The electronic communications and rocket technologies required to develop satellite communications all had their roots in World War II. The satellites themselves had a more complicated origin. Most space vehicles had been envisioned as manned vehicles until shortly before the launch of *Sputnik 1* by the Soviet Union in 1957. One could argue that some people—in Congress and at NASA—still think of space vehicles as manned. The aircraft industry had experience building manned high-altitude vehicles, but the problems of space were different. Unmanned satellites would be built by the electronics industry—not the aircraft industry. The war provided the first real launch vehicle, the German V-2. This launch vehicle could not put a satellite into orbit, but it could penetrate the atmosphere and sample the space environment. Ineffective as a weapon, the V-2 provided a major impetus for the "space cadets." All of their concepts and visions were at least partially validated by the production of a large rocket. The "space cadets" believed that just a little more effort would produce a satellite launch vehicle—but who would pay for the development?[2]

The Economy and National Security

The U.S. economy boomed after the war—albeit somewhat erratically with periodic recessions (see Figure 2.1). The returning U.S. veterans were eager to catch up. The Serviceman's Readjustment Act of 1944 provided for educational grants and home loans. Many veterans got married, bought a house, had children, and went to college—almost simultaneously. The "can do" and "hurry up" attitudes fostered by the war were now being applied to normal life. The entire nation had cooperated to defeat fascism. There was a feeling that Americans could do anything. World War II had lifted the United States out of the Great Depression. The New Deal may have improved morale and ameliorated the worst effects of the depression, and industry may have recovered somewhat by 1936–37, but it took the war to put U.S. factories back on multiple shifts.

Gross National Product (GNP) grew by more than 5 percent per annum from 1940 to 1944 and by more than 10 percent per annum from 1941 to 1943. The peak growth rate in 1941 was over 15 percent. By the end of the war, GNP per capita was almost twice what it had been at the beginning of the war. The postwar recession (especially 1946) was painful to many, but it was milder than

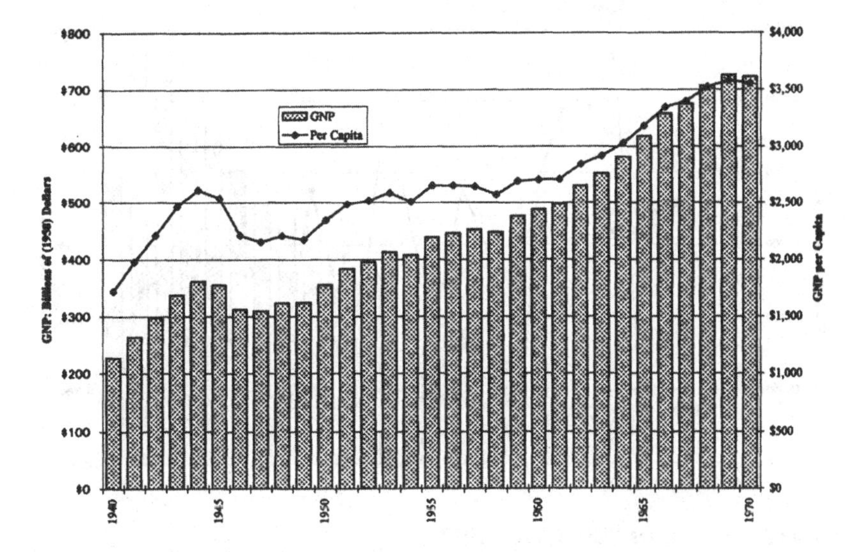

Figure 2.1. U.S. GNP Growth, 1940–1970.

some had feared. The savings accumulated during the war years ensured that consumption grew even during the recession. The Korean War rekindled the economic fires for a few years in the 1950s, but growth was erratic. The continuous stable growth of the 1960s was to be unique.

In many ways World War II was an electronic war. Radar may have been the most significant technology the Allies had, especially in the early defensive stage of the war. Other electronic technologies included the VT (variable time: proximity) fuse, fire control systems (and other primitive computers), and the global communications systems required to coordinate global war. There were great hopes that these new technologies would have a place in the postwar civilian world. Radio had kept the home front informed, but television, a technological innovation delayed by the war, was seen as the next step forward in communications technology. The words of Erik Barnouw in *Tube of Plenty* probably best express the television boom:

> In 1945, as peace came, it was possible to discern an explosive set of circumstances.
>
> Electronic assembly lines, freed from the production of electronic war matériel, were ready to turn out picture tubes and television sets. Consumers, long confronted by wartime shortages and rationing, had accumulated savings and were ready to buy. Manufacturers of many kinds were eager to advertise. The situation awaited a catalyst, a signal. It came with surprising suddenness.

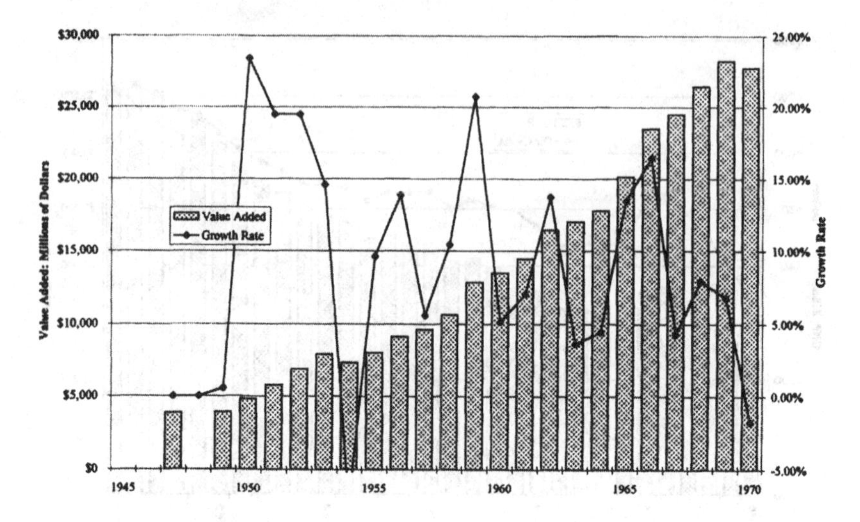

Figure 2.2. Electronics Market, 1940–1970.

> In 1945, the FCC [Federal Communications Commission], once more reviewing spectrum allocations, made crucial decisions. It decided to resume television licensing. . . .
>
> The pace of television activity quickened. By July 1946 the FCC had issued twenty-four new licenses. Returning servicemen with radar experience, whose knowledge was convertible to television, were snapped up by many stations. Advertising agencies were ready; many had already formed television departments and had experimented with television commercials and programming.[3]

In the summer of 1946, RCA got its black-and-white sets on the market.

Although the electronics market grew dramatically after World War II, much of it was subject to erratic ups and downs as the military market gained and lost importance (see Figure 2.2). Many new firms joined the older radio companies to share the government largesse: Raytheon, Litton, and Hughes Aircraft Company. The last, despite its name, was an electronics company. All of these companies were, to varying degrees, subject to the whims of congressional budgets for military electronics. The capability to build electronic devices existed, but it was endangered by erratic military budgets. What was needed was a secure commercial market. Television was one answer to this problem—transatlantic communications might be another.[4]

If the electronics industry could be characterized as successful overall, with a few ups and downs in military electronics, the aircraft industry was almost all downs (see Figure 2.3). President Franklin D. Roosevelt's pledge to build more

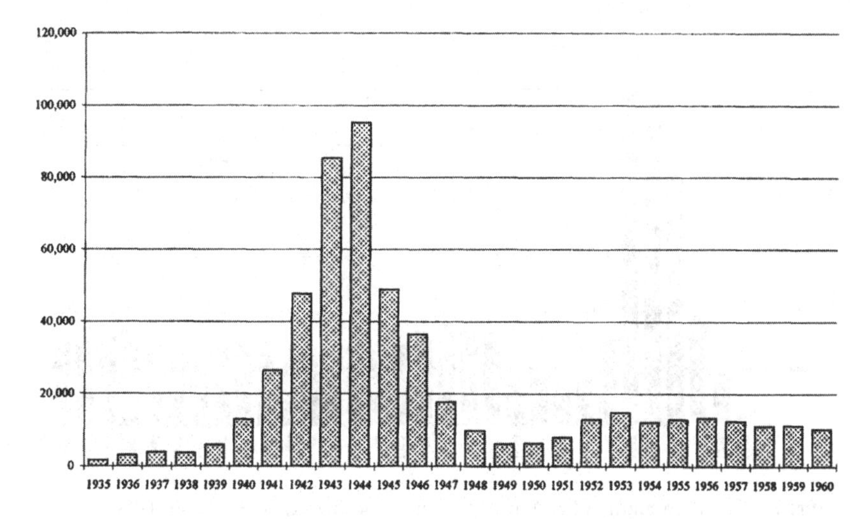

Figure 2.3. Aircraft Manufactured in the United States, 1935–1960.

airplanes than ever before was clearly met during the war. At its peak in 1944 the U.S. aircraft industry was manufacturing almost 100,000 airplanes each year. Although the Air Force was to receive favored treatment in the postwar years, production never again approached even 20,000 airplanes per year. Lockheed President Robert Gross commented that as long as he lived, he would never forget "those short appalling weeks," after the surrender of Japan, when the war orders stopped.[5] The aircraft manufacturers might have gone bankrupt without peacetime government orders. The cold war and the Korean War got them back on their feet, but the glory days and high profits of World War II were over. Only Boeing and Douglas had continuing success in the commercial airplane market. Lockheed built excellent airplanes, but only the Constellation was really successful in the commercial market. For the rest of the industry, military orders were all that kept the doors open.

The postwar emphasis on nuclear weapons and strategic power gave Boeing a large bomber market (B-29, B-50, B-47, and B-52), which provided the technology and capital that eventually allowed it to capture the commercial jet airliner market. In 1945, with the end of the huge war orders, Boeing's president, Phil Johnson, was prepared to change to other fields, including automobiles and furniture.[6] The success of Boeing bombers and, eventually, civilian aircraft made these drastic business changes unnecessary, but Boeing, like many other airplane manufacturers, began looking into missiles.

The cold war began even before the "hot war" ended, but nothing could stop the country's desire to get the soldiers home and do what America had done

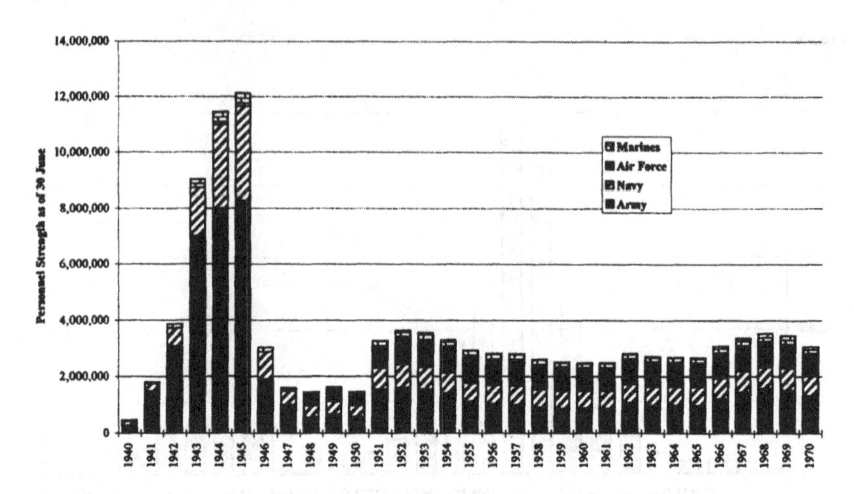

Figure 2.4. Total Strength of the Armed Forces of the United States, 1940–1970.

after every previous war: disarm. As late as 1940, the United States had less than 1 million men under arms (see Figure 2.4). By 1945, the United States had 12 million men under arms. After the Japanese surrender, troop strength was rapidly reduced to about 3 million in 1946. From 1947 to 1950, the United States had less than 2 million men under arms. The Korean War almost doubled this number, but even then, U.S. forces remained well under 4 million. In the immediate postwar period, the Soviet Union lowered the iron curtain, which closed off Eastern Europe from Western Europe. Communist parties gained control of Yugoslavia, Hungary, Poland, Czechoslovakia, Romania, Bulgaria, Albania, and East Germany. Postwar civil-political strife, especially in Greece, caused President Harry Truman (on March 12, 1947) to propose military and economic aid to any country facing communist subversion. The Berlin blockade (July 24, 1948, to September 30, 1949) was seen as one more step in the Soviet quest for European conquest. In the midst of these East-West conflicts in Europe, Truman was informed (in September 1949) that the Soviet Union had detonated an atomic bomb of its own and that the Nationalist Chinese forces had evacuated to Taiwan as Mao Tsetung's forces finally prevailed.[7]

At home, where obscurantist anticommunism was preparing the way for McCarthyism, the military establishment was at war with itself. The "revolt of the admirals" made painfully obvious the perpetual war between the services. The battle against the logical unification of the armed services was fought, successfully, by James V. Forrestal. As secretary of the Navy, he defended the Navy's claim that it was so different that no Army or Air Force officer could ever give direction to its units. The National Security Act of 1947, instead of unifying the services, merely provided unified management at the cabinet level. Forrestal was rewarded—or punished—by being named the first secretary of defense. The seventy-wing Air Force had momentarily beaten the Navy, but the admirals soon got their carriers.

Conspiracy theories have suggested that the cold war was simply a device to save the aircraft manufacturers from bankruptcy.[8] Whatever the original cause, the invasion of South Korea on June 25, 1950, established national security as the most important political issue. Before the Korean conflict, the United States had disarmed after every war. Since the Korean War, the United States has maintained the most powerful armed forces in the world. In the process, science and technology policy came to be dominated by national security concerns.

Rockets, Satellites, and the Military

In late 1945 the U.S. Navy began an effort aimed at developing launch vehicles and satellites.[9] Aerojet, the commercial cousin of the Guggenheim Aeronautical Laboratory of the California Institute of Technology (GALCIT), and the Jet Propulsion Laboratory (JPL), the Army cousin of GALCIT, had designed and tested—under Navy auspices—a hydrogen-oxygen rocket motor that could be used in a satellite launch vehicle. The Navy Committee for Evaluating the Feasibility of Space Rocketry (CEFSR) contracted with GALCIT to perform research on the expected performance parameters of the desired rocket and satellite.

It quickly became obvious that Navy support for development would not be forthcoming. Members of CEFSR approached the U.S. Army Air Force (AAF) and suggested that a joint program be launched. Initial reaction was extremely favorable, but ultimately the AAF was as unwilling as the Navy to fund the project. One result of this interaction was that the AAF asked the Project RAND organization to perform an independent study of the feasibility of an Earth satellite.[10] The study was completed on May 12, 1946.[11] The first RAND report made many observations, but four have special meaning: (1) a 500-pound satellite could be placed into a 300-mile orbit within five years (i.e., 1951); (2) a four-stage vehicle would be required, and a single-stage-to-orbit (SSTO) vehicle—the Navy design—would be impossible; (3) a satellite launch would have a dramatic effect on world opinion; and (4) the satellite might have significant use as a *communications relay station*.

A second RAND study was begun immediately; its report was published in February 1947.[12] James E. Lipp, a CalTech Ph.D., was responsible for this study and wrote some of the conclusions. Lipp pointed out the advantages of low polar orbits for reconnaissance satellites and the advantages of twenty-four-hour equatorial orbits for communications satellites. This study emphasized needed research on subsystems and technologies for space flight. But the timing was unfortunate: in December 1946 the Truman administration had ordered a curtailment of military R&D activities. This resulted in a general reduction in funds for high technology, especially leading-edge technologies with no immediate payback. Two years later, in July 1949, another RAND study was published, elaborating the idea of a geosynchronous communication satellite.[13]

The Air Force resumed its space studies in earnest in 1949. A 1949–50 RAND study emphasized two points: first, whichever nation first launched a satellite would score a huge psychological blow, and second, the satellite could be used as a reconnaissance vehicle.[14] According to the historian Walter McDougall, "The RAND document of October 1950, more than any other, deserves to be considered the birth certificate of American Space Policy."[15] Work on a reconnaissance vehicle proceeded, at the study level, for the next few years at RAND, the Air Force, North American Aviation, Lockheed, RCA, and other organizations.

Military interest in missiles and rockets continued to suffer ups and downs. Although the German V-2 had made rockets respectable after 1944, during the prewar years "rocket fever" had attracted as many showmen as scientists. Some of the stunts and the more sensationalistic science fiction had contributed to a bad image for rockets. The Jet Propulsion Laboratory was actually a *rocket* propulsion laboratory, for example, but the term was not respectable enough for CalTech and the Army. The world of rocketry was filled with an uneven mix of "space cadets" (the space enthusiasts who had given rockets a bad name in the 1920s and 1930s), weapons builders, scientists, and engineers. All contributed to building the rockets that made satellite communications possible.

The American Interplanetary Society (AIS) was formed on April 4, 1930, in the home of G. Edward Pendray and eleven other science-fiction writers and fans. In spite of its "space cadet" orientation, the AIS, like its British counterpart, the BIS, sponsored serious research. Robert H. Goddard, the American rocket pioneer, became a member but was never active in the society. The AIS built at least five rockets but then concentrated on static firing of engine designs—including James H. Wyld's regeneratively cooled design. In 1938 the AIS changed its name to the American Rocket Society (ARS) in the hope of improving its image. In December 1941 members of the ARS formed Reaction Motors Inc. (RMI) to seek military contracts using the Wyld design.[16]

GALCIT meanwhile had begun its studies of rockets in 1936 when Frank Malina, a CalTech graduate student, convinced Clark Millikan, his adviser and the son of the CalTech president, Robert Millikan, that a doctoral dissertation on the problems of rocket propulsion was a suitable contribution to the field of aeronautics. In 1936 and 1937 Malina tested rocket engines in nearby Arroyo Seco. Malina presented the results at a meeting of the Institute of Aeronautical Sciences (IAS) in January 1938. The resulting publicity emphasized manned Moon flights, but Malina and GALCIT stressed that their ultimate goal was to build a sounding rocket that could reach an altitude of 30 kilometers. The approaching war brought contracts for jet-assisted takeoff (JATO) units—both liquid- and solid-fueled. In 1942 the senior members of the GALCIT rocket team formed their own company: Aerojet. In 1943 the Army suggested that GALCIT look into building a long-range bombardment missile. JPL was formed as a federally funded CalTech research center to execute this program. By 1945 the JPL-built, liquid-fueled WAC Corporal had been launched to an altitude of 40 kilometers.[17]

By the late 1940s, U.S. rocket technology was probably more advanced than German technology had been at the end of the war. The missing element was large-scale production of large rockets. Only the Germans had done this. The problems of satellites had been evaluated, but commitment was slower in coming. Rockets were seen as a potentially "war-winning" weapon. The eventual success of rockets was ensured, but first they had to compete with manned bombers.

The bomber forces of the United States were not used against the Japanese main islands until late in the war because no base was available from which to launch them. The B-17 and its successor, the B-29, were not capable—in spite of their names ("Flying Fortress" and "Superfortress")—of defending themselves against modern fighter aircraft. The P-47 and the P-51 escort fighters made late World War II strategic bombing much more effective than it had been. The postwar B-50 (an improved B-29) and B-36 were similarly incapable of defending themselves. Modern propeller-driven fighters were capable of speeds well over 400 knots by the end of the war. No production-model propeller-driven bomber could reach 400 knots. Cruise speeds were typically 250 knots. The B-29, B-50, and B-36 did have very long range capability, but only the huge B-36 could actually reach Soviet targets from the United States. No fighter aircraft were capable of reaching Soviet targets from U.S. airfields. The B-36 had some hope of flying high enough to escape propeller-driven fighters, but the advent of modern jet aircraft was to make it increasingly vulnerable.

Both nuclear weapons and their delivery systems were controversial in the postwar period. The weapons controversies were eventually settled in favor of building the maximum number of the most powerful weapons[18] and putting them on every conceivable delivery system. The Air Force was committed to the winged, manned bomber as a delivery system, but the German V-1 and V-2 had planted the notion that missiles could not be stopped. The V-1 was actually not very difficult to stop with radar warning and other interception technologies, but the V-2 was truly unstoppable.

A series of studies by RAND and Theodore von Karman stressed the possibilities inherent in missiles—including their ability to put satellites into Earth orbit. Von Karman, in *Toward New Horizons* (1945), his aerospace technology analysis performed for General Henry "Hap" Arnold, was sufficiently sanguine about intercontinental missiles to counteract Vannevar Bush's view that these missiles were "futuristic."[19] In 1946 the Air Force had a variety of missile programs in the works; among these were several surface-to-surface (bombardment) missiles: North American Aviation's Navaho (MX-770), Glenn L. Martin Company's Matador (MX-771), Northrop's Snark (MX-775), and Convair's Atlas (MX-774). Only the Navaho and the Atlas were rockets; only the Atlas was a ballistic missile—it had no wings. The Navaho is particularly significant: North American Aviation's engineers had been able to inspect V-2 engines before the end of the war, and based on what they learned, North American then produced its own prototype rocket motor. The Rocketdyne division of North American, with the later assistance of some of the German V-2 engineers, was thus able to mass-produce relatively large rocket engines for Navaho. Although Navaho was a failure, its engines, continuously improved over many years, were to be used on the Atlas and Thor missiles.

Perhaps indicative of the Air Force attitude (which advocated the primacy of pilots) was that bomber designations (the Atlas was the XB-65) were given to surface-to-surface missiles and fighter designations were given to air-to-air missiles. Atlas was among the projects canceled in 1947. By 1950 only the Navaho and Snark surface-to-surface missiles were still funded. A RAND study that year told the Air Force that long-range nuclear-tipped missiles were now feasible.[20] Convair had continued the MX-774 program after cancellation, using internal funding and the remaining Air Force funding—and it eventually launched an MX-774 successfully. In the interim, studies had continued. When the Air Force awarded a study contract, Convair was ready. On January 23, 1951, Project MX-1593 was begun to develop a ballistic missile capable of carrying an 8,000-pound atomic warhead 5,000 nautical miles and striking within 1,500 feet of its target. The total development budget was approximately one quarter of a billion dollars spread over the following dozen years.[21] In 1954 the Tea Cup Committee (TCP, Technological Capabilities Panel), headed by John von Neuman, and a RAND report both recommended that the Atlas development be accelerated on a "crash" basis, that the missile be capable of carrying a 1,500-pound warhead rather than an 8,000-pound warhead, and that the accuracy requirement be reduced to 3 miles rather than 1,500 feet.[22] Technology had made nuclear weapons both smaller and more powerful. The first Atlas squadron was organized in 1958. In 1961 operational status was achieved.

The Thor IRBM program had multiple arguments to justify its existence. It was seen as a less risky, quicker development than the Atlas, and it would use many of the same components (including a rocket motor from Rocketdyne). It was seen as a replacement for the Matador tactical missile. Last, but not least, it was seen as a weapon system that, like the Army's Jupiter, could be deployed in friendly European countries within IRBM range of the Soviet Union. Begun in 1955, operational Thor missiles were handed over to the Royal Air Force (RAF) in 1960.

By the late 1950s a variety of rockets were becoming available. To the 1940s-vintage Army Redstone bombardment missile and the Navy Viking sounding rocket (basically an American V-2) were added the newer Army Jupiter IRBM (a Redstone successor), the Navy Polaris IRBM, the Air Force Thor IRBM, and the Air Force Atlas ICBM (intercontinental ballistic missile). Already in the advanced design stage were the Titan ICBM and the solid-fueled Minuteman ICBM.

Space Cadets and the IGY

By 1945 U.S. rocketry had a theoretical basis and an experimental basis. Three commercial firms—North American Aviation, Aerojet, and RMI—were building liquid-fueled rocket engines, and several companies were mixing solid fuel. The WAC Corporal was the largest American rocket—at 300 kilograms much smaller than the 13,000-kilogram V-2. It did have certain unique characteristics: its liquid propellants, aniline and nitric acid, were storable and hypergolic (self-igniting). As part of the spoils of war, the United States acquired a large number of V-2 components and most of the senior technical personnel from the German V-2 R&D facility at Peenemunde. In late 1945 the V-2 components went to White Sands, New Mexico, and the German rocket team went to nearby Fort Bliss. The goal of the Army's Project Hermes and its contractor, General Electric, was to build a tactical bombardment missile. Toward that end, they expected to learn how to assemble and launch the German V-2s with the assistance of their German "rocket team."[23]

Colonel Holger N. Toftoy, the Army ordnance officer responsible for the V-2 launches, decided to make the launches available to the scientific community. The Navy, which had no missiles, had a functioning scientific team through its Office of Research and Inventions (ORI), its Naval Research Laboratory (NRL), and its contractor the Applied Physics Laboratory (APL) of Johns Hopkins University. The scientists were grouped together as the V-2 Panel (later the Upper Atmosphere Rocket Research Panel—UARRP). Starting in 1946, they began placing scientific payloads on Army V-2s. In 1946 the WAC Corporal was improved and converted into the Aerobee (a combination of the words "Aerojet" and "Bumblebee," the Navy's project) sounding rocket. Because of its low cost (about $20,000) and reasonable performance (70 kilograms could be lofted to 130 kilometers), by 1949 the Aerobee was the most-used sounding rocket. Meanwhile, the Navy had determined to build its own sounding rocket to replace the V-2. Originally named the Neptune, the (later-named) Viking used a rocket motor built by RMI, a known quantity. The airframe would be built by Glenn L. Martin Company, which would also integrate the final package. Though the Viking was more sophisticated than the V-2 and more efficient in many ways, only a few were built and fired from the late 1940s to the mid-1950s—in spite of the relatively low cost ($200,00) and good performance (70 kilograms could be lofted to 400 kilometers).[24]

White Sands saw a collaboration among the military, scientists, and the German rocket team. Several members of these groups were "space cadets"—dedicated to the conquest of space. This group included Fred Whipple, the Harvard astronomer who had participated in the early RAND studies, and Wernher von Braun, the principal space cadet and the head of the German rocket team. Von Braun and his colleagues had been disappointed that their participation in the White Sands activities was only as advisers to the Army-General Electric team. They were hoping to do more innovative work. To while away the time, they worked on a study for a mission to Mars. This was published in German in 1949 as *Das Mars Projekt* and in 1952 was translated into English as *The Mars Project*.[25] In January 1947 von Braun gave a speech—"The Future Development of the Rocket"—to the El Paso, Texas, Rotary Club.[26] In 1950 von Braun and his team were moved from Fort Bliss, Texas, to the Redstone Arsenal in Huntsville, Alabama. Shortly thereafter, Willy Ley, one of the original prewar space cadets, was able to organize the First Annual Symposium on Space Travel at the New York Museum of Natural History on October 12, 1951. This was followed by a conference on the physics and medicine of the upper atmosphere, held in San Antonio, Texas. Von Braun did not speak at either meeting, but in discussions outside of the meeting he so impressed Cornelius Ryan of *Collier's* magazine that Ryan decided to do a series of articles on space travel.

Ryan obtained the support of the conference attendees, especially Whipple, James van Allen, and von Braun. He also enlisted several technical artists: Chesley Bonestell, Fred Freeman, and Rolf Klep. The first space issue, with an issue date of March 22, 1952, covered the details of building rockets and a space station in Earth orbit. The public response was extremely enthusiastic. *Collier's* followed with two issues, on October 18 and October 25, 1952, detailing a manned expedition to the Moon. Three later issues, on February 28, March 7, and March

14, 1953, discussed the physiology of space travel. The June 27, 1953, issue of *Collier's* described the launch of an unmanned space station. The last space issue, on April 30, 1954, described a manned expedition to Mars. Book-length descriptions of the conquest of space had appeared in 1949 (Willy Ley and Chesley Bonestell's *Conquest of Space*) and 1951 (Arthur C. Clarke's *Exploration of Space*), but the *Collier's* articles reached a much larger audience. The publicity blitz surrounding the articles made a media star of von Braun. The articles were later expanded into books by Ryan[27] and into Walt Disney television productions entitled "Man in Space" and "Man and the Moon" in 1955. In December 1957, Disney aired "Mars and Beyond."[28]

Just before the Hayden planetarium meeting in New York, on September 4, 1951, the International Astronautical Federation (IAF) was founded in London by members of the BIS, Gesselschaft für Weltraumforschung (GfW, successor to the VfR), ARS, and other rocket and interplanetary societies. A preliminary meeting had been held the previous year in Paris. Arthur C. Clarke, as chairman of the BIS, hosted the meeting. Eugen Sänger was elected president of the IAF. The theme of the technical presentations in London was "The Artificial Satellite." The members of the IAF knew that the conquest of space was coming. The basic technologies already existed—only money and commitment were required.[29]

The U.S. Navy, having ended its satellite research in 1948, began again in 1954 with two actions: it experimented with the Moon as a passive communications satellite, and it proposed to the U.S. Army a joint satellite endeavor: Project Orbiter. Two different events appear to have precipitated Project Orbiter. First was the series of activities associated with the International Geophysical Year (IGY). Originally proposed as an alternative to another International Polar Year, the IGY would investigate a broad range of terrestrial (and solar) phenomena. Throughout 1952–54 the ARS suggested that an Earth satellite would be an appropriate part of the IGY. On October 4, 1954, the IGY committee (CSAGI) suggested that a small scientific satellite be placed in Earth orbit to study the Sun and the upper atmosphere. Second, on March 1, 1954, RAND had delivered the Project Feedback final report recommending that development of a reconnaissance satellite be a top defense priority. On March 16, 1955, the Air Force briefed industry on the requirements for Weapons System WS-117L, a "strategic satellite system." The Air Force had a mission in space; the Army and Navy wanted a mission too.[30]

The Army had a potential launch vehicle, the Redstone rocket, and at the Office of Naval Research (ONR), the Navy had a group that had been building sounding rocket science payloads for several years. Navy Commander George Hoover was to direct the project. By March 1955, Donald Quarles, assistant secretary of defense for R&D, had two proposals in hand: Project Orbiter, sponsored by James Smith, assistant secretary of the Navy for air; and a proposal from the NRL to use a variant of the Navy's Viking sounding rocket, built by Glenn L. Martin Company. A committee, headed by Homer J. Stewart of JPL, examined the two proposals and eventually recommended that the NRL proposal be pursued. Three launch vehicles were considered: Atlas, Redstone, and the upgraded Viking (Vanguard). Atlas was rejected as too important to national security to be part of a science project. The Redstone had limited capability (it was improved by January 31, 1958).[31]

The Communications and Electronics Industries

World War II saw vast advances in electronics, especially high-frequency (microwave) devices for radar, which led to a postwar television boom, initially in the United States and somewhat later in Europe, which in fact had television before the war. Electronic computers were also developed during the war—none small, some requiring huge buildings full of electronic and electro-mechanical apparatus. To a far greater degree than World War I, World War II was truly global. In the words of the economic historian Alan Milward: "There were few underdeveloped areas of the globe in which the relatively well-fed and well-paid armed men of the combatant powers did not descend with their technology and their money."[32] The "global village" may not have been realized until the late 1960s, but its origins are visible in World War II. Communications in its broadest sense—film, print, radio, and especially television—brought the world together, for better or worse.

The American people had experienced several years of forced savings and were ready to buy goods of all varieties. Manufacturers were eager to convert back to civilian production and to advertise their new wares. The United States had not truly entered into the "television age" before World War II, as many European nations had. It was not until 1941 that the FCC agreed on a television standard; immediately after the end of the war, it resumed licensing television stations.

The end of World War II brought about a dramatic increase in telephone use. Communications grew at more than three times the GNP rate of growth. A large component of this use was long-distance—both domestic and international. International trade was increasing unevenly but showed great promise. This growth in international trade was fueled by the status of the United States as the only major industrial power not to have had its industrial plant devastated during the war. That industrial plant had, however, suffered somewhat due to overuse, and it was wearing out. The overall health of the U.S. economy suggested that rebuilding the civilian side of the U.S. industrial might, static since 1941—in some cases static since 1929—would be a relatively easy task.

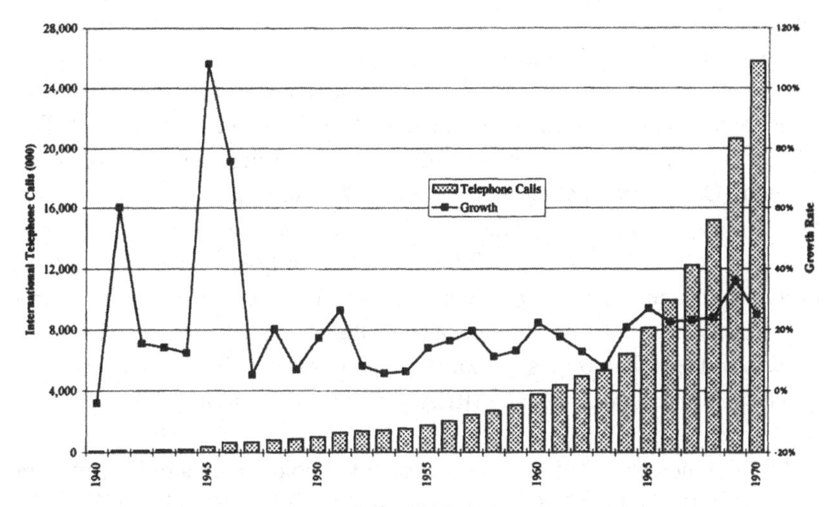

Figure 2.5. International Telephone Calls, 1940–1970.

Microwave[33] relay systems had been used by the U.S. Army Signal Corps in the European and Pacific theaters. Their use for domestic long-distance telephony was seen as an obvious application. Several companies—including Philco, RCA, GE, Raytheon, Western Union, and IBM—were interested in building a microwave relay system for telephony and/or video relay. On April 18, 1945, Philco demonstrated a television relay from Washington, D.C., to Philadelphia, Pennsylvania. The FCC, although it encouraged experimentation in the microwave frequencies, was not ready to make the Bell System's investment in copper wire obsolete overnight. AT&T continued to use copper wire and began to implement a coaxial cable system to provide long-distance telephony. By the late 1950s, after a 1949 antitrust suit had been settled by the 1956 Consent Decree, AT&T was willing to allow others (e.g., Western Union) to build microwave relay systems.[34]

International telecommunications was in a strange state: very little improvement had been made since the last telegraph cable had been laid in 1928. Radio-telegraphy and radio-telephony, though rapidly overtaking cable telegraphy, suffered from the "noise" associated with the technology (HF/AM) then in use. Improvements in radio during the war were applied to civilian commercial use shortly thereafter (see Figure 2.5). The biggest improvement was the laying, by AT&T and its European partners, of TAT-1, the first Trans-Atlantic Telephone (coaxial) cable in 1956. Although radio-telephony had been demonstrated in 1915 and provided commercially since 1927, this was the first time that telephone calls could be relayed across the Atlantic without regard to weather or sunspot cycles. Even more miraculously, individual voices could be recognized consistently—tone and nuance used in normal conversation could be transmitted.

TAT-1 was primarily an AT&T venture, but since it was an international cable, foreign correspondents were needed. AT&T and the British Post Office (BPO) were equal owners, with Canada's Overseas Telecommunications Corporation (OTC) participating at a lower level. The major part of the cable, Scotland to Newfoundland, was laid by AT&T using Western Electric repeaters. The total cost of TAT-1 was $50 million. Only thirty-six voice circuits (i.e., thirty-six channels or half-circuits in each direction) were available. Later improvements in technology would increase this number to seventy-four. The success of the technology is indicated by the immediate inclusion of the telegraph cable and radio companies (RCA, ITT, and Western Union) as part of the TAT arrangements.[35]

TAT-2 followed in 1959, using the same technology but laid directly from England to the United States. The owners of this cable included AT&T, the French Post, Telegraph, and Telephone (PTT) organization, and Deutsche Bundespost (DBP). The cost was slightly lower: only $43 million.[36] TAT-3 was another AT&T-BPO cable, laid in 1963. It cost $51 million but used new technology providing 138 basic voice circuits: more than three times the initial TAT-1 capability.[37] TAT-4 was laid in 1965 amid some discussion of its cost-effectiveness relative to satellites. TAT-5, laid in 1970, provided a dramatic increase in capability—to 720 voice circuits—for a relatively modest $79 million.

The advances in communications technology in World War I, combined with the U.S. view of cable communications as a British monopoly and the widely held basic radio patents, led to the formation of the Radio Corporation of America (RCA) as a U.S. patent pool to break the British cable and radio (Marconi) monopolies and led also, serendipitously, to the birth of radio broadcasting. When RCA was formed in 1919–20, part of the rationale was to pool the radio patents held by many separate companies in order to allow commercial exploitation of the World War I advances made using all available patents. AT&T had developed the first commercially successful vacuum tube. GE had a high-frequency alternator and, because of its lightbulb facilities, was a major manufacturer of vacuum tubes. Westinghouse also owned patents, as did smaller players like the MacKays. The final allocation of RCA ownership was as follows: GE, 30.1%; Westinghouse, 20.6%; AT&T, 10.3%; United Fruit, 4.1%; and individuals, 34.9%. AT&T also received (or thought it received) a monopoly on the use of the patents for telephony—similar to the RCA monopoly on the use of the patents for radio.[38]

Although RCA made many of the technological developments in radio and television, AT&T, specifically Bell Labs, made many of the basic electronic device breakthroughs. The most famous of these was the invention of the transistor in 1947, but AT&T was also deeply involved in the development of solar cells, the MASER and the TWT (traveling-wave tube). The MASER is a very low noise amplifier—important for Earth stations receiving weak satellite signals—and the TWT is a very high gain linear amplifier providing a light-weight single-stage amplifier for the satellite transmitter.

AT&T had been a perennial target for anti-trust suits. One of the first of these was due to its reorganization by J. P. Morgan in 1907, incorporating the Western Union Telegraph Company. In 1913 AT&T was forced to divest itself of Western Union and promise to cease its monopolistic practice of buying up small phone companies. AT&T may have been glad to give up the Western Union telegraph network, but it kept the old Western Union manufacturing arm—Western Electric—which it had acquired in the late 1870s after winning the telephone patent battle. AT&T had several organized research efforts dating to the years when Bell's original patents had begun to expire at the end of the nineteenth century. In its final form, the major research effort was centralized in Western Electric's Bell Telephone Laboratories (BTL). BTL was responsible for many great discoveries, notably the previously mentioned transistor, as well as radio astronomy and the 2.7K cosmic background radiation; all these efforts were directed toward the betterment of AT&T. MASERs may have originated at Columbia, but AT&T-BTL worked hard at perfecting the technology for its submarine telephone cables. The TWT was developed at BTL and used for microwave relay stations.

In a measure of its regard for AT&T, the U.S. government drafted the company to perform nuclear weapons work (Sandia), missile work (Safeguard), and systems engineering support for the Apollo program (BellComm). Other companies also had R&D laboratories, but AT&T stood at the top in terms of the caliber of the R&D performed and the willingness of the company to invest vast sums. All of those who would develop the early TWTs for communications satellites had, at some point, worked at Bell Labs with John Pierce and his team.

Most of the companies that would eventually be involved in satellite communications had been around for some time. General Electric (GE) was formed in the merger of Thompson-Houston and Edison General Electric in 1892. AT&T was formed in 1885, first as a long-distance provider and then as a general manager for the Bell Telephone operating companies. ITT and RCA were formed in the 1920s. ITT was established to manage a variety of telephone and telegraph companies, initially in Latin America and later in Europe when AT&T was forced to dispose of its international holdings. GTE was a domestic competitor of AT&T. Hughes was formed in the 1930s as a hobby-shop for Howard Hughes—it grew dramatically in the postwar period. All of these companies

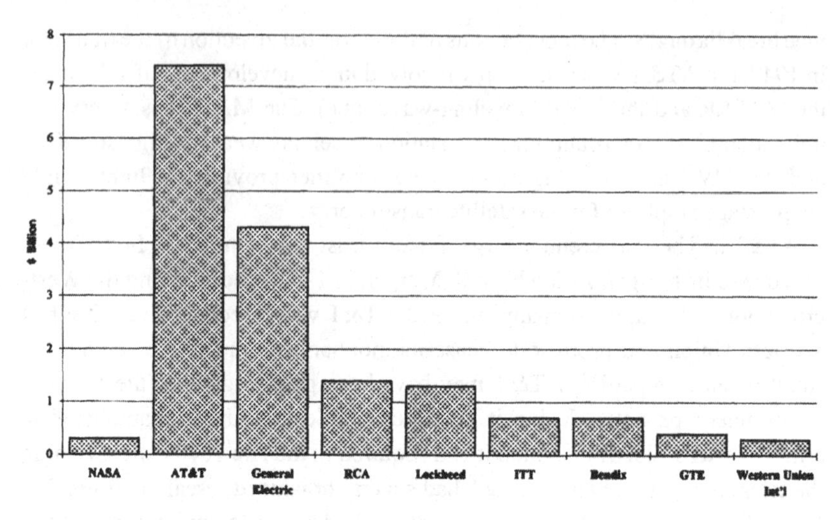

Figure 2.6. Industry Sales and NASA Budget, 1959.

were major electronics manufacturers. All except Hughes and GE had significant stakes as communications system operators.

These were generally quite large companies. The chart in Figure 2.6 provides a relative measure of size. All figures are in billions of current (1959) dollars. Industry figures are net sales. During this period, U.S. GNP was $483.7 billion, federal outlays were $92.1 billion, and defense spending was $44.6 billion. In 1959 all of these companies outspent NASA. AT&T and GE alone accounted for over 2 percent of GNP. AT&T was concentrated in telecommunications. It dominated its industry as no other company did. Any improvement in telecommunications technology was ultimately to AT&T's advantage.

Visionaries and Prophets

Rockets seem to have dominated early space studies far more than satellites. The rationale is fairly obvious: without rockets, satellites are impossible; rockets must come before satellites. Nevertheless, many analysts envisioned uses for rocket-launched satellites. Curiously, at least in retrospect, most of the hypothesized satellites in the postwar period were space stations—*manned* satellites. This mindset was inherited, at least in part, from the air forces, where it was assumed that "flight" vehicles required pilots. Only the space cadets talked about unmanned satellites.

The acknowledged godfather of communications satellites, Arthur C. Clarke, was an RAF officer in 1945; he hoped to revitalize the BIS after the war. In October 1945 his article, "Extraterrestrial Relays," was printed in *Wireless World*.[39]

Although Herman Oberth in 1923 and George O. Smith in 1942 had theorized about geosynchronous or, in Smith's case, Earth-Venus relay communications satellites, Clarke was the first to develop more fully the concept of a satellite that orbited the Earth in twenty-four hours—the satellite would rotate at the same rate as the Earth and thus appear stationary. Three of these geosynchronous satellites, each fixed over a specific longitude on the equator, would cover the entire globe. The satellites would be used for broadcasting—especially television distribution. Several of Clarke's assumptions turned out to be false—or at least premature. Clarke assumed the station would be manned because the vacuum tubes would have to be changed on a regular basis. He also assumed the three basic locations would be over land masses—rather than over oceans. Since the original article was not well publicized, Clarke's real contribution was to continue pressing for the geosynchronous system. His 1951-52 book, *The Exploration of Space,* was viewed by many as a blueprint for the entire space program.[40] In that book he also included the concept of global communications using three geosynchronous satellites. In Homer Newell's retrospective on his NASA career, he credits Clarke as the champion of communications satellite applications.[41] Clarke even referred to himself as the "godfather" of satellite communications,[42] on the grounds that he had very little to do with the actual technical developments or the implementation.

Clarke's satellites would have been huge—weighing hundreds of tons rather than hundreds of kilograms. Their purpose, international television broadcasting, might be politically impossible. Clarke powered his satellites with solar steam boilers but imagined solar-electric devices in the near future. He mentioned nuclear power only as a propulsion source.

In 1949 another member of the BIS, Eric Burgess, published an article on communications satellites.[43] Burgess analyzed the GEO (geosynchronous Earth orbit) in some detail, paying special attention to the various perturbations and to the type of rockets required to reach the 42,000-kilometer orbit (36,000 kilometers above the surface of the Earth).[44] He also described a possible satellite configuration in detail. Like Clarke, he expected broadcast television and radio to be the major justification for the geosynchronous satellite, but unlike Clarke, he expected the satellite to be unmanned, at least temporarily, until manned stations could be constructed. The proposed satellite was body-stabilized using gyros and included the final propulsion stage. Like Clarke, Burgess proposed to use a solar steam boiler to generate power. The total mass of his satellite was 600 kilograms ("dry"—the fuel added another 1,200 kilograms). He predicted the satellite would be launched by 1959.

Science fiction was John Robinson Pierce's introduction to electronics and space. After earning a Ph.D. at CalTech, Pierce became a senior scientist-

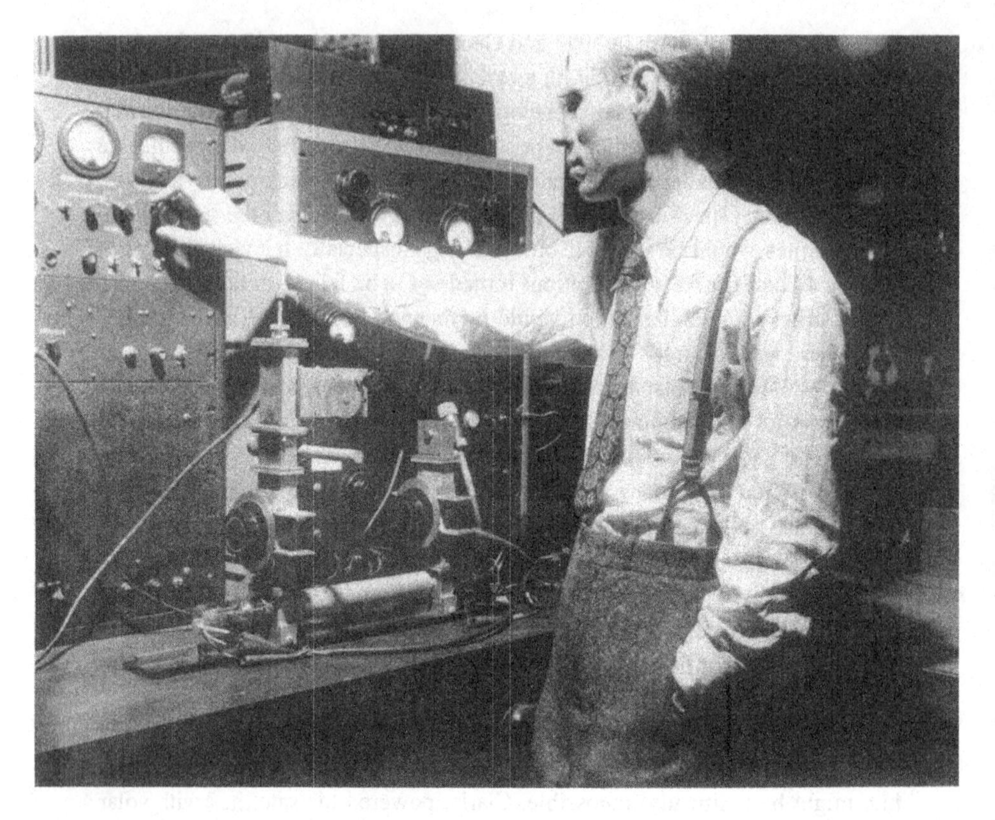

John R. Pierce testing a traveling-wave tube, 1949. (Courtesy AT&T)

engineer-manager at AT&T's Bell Telephone Laboratories. Pierce's interest in science fiction led him to write an article on space communications—"Don't Write; Telegraph"—for the March 1952 issue of *Astounding Science Fiction.* Written under his pseudonym, J. J. Coupling, the article was influenced by Pierce's reading of George O. Smith's article, "QRM Interplanetary," in the October 1942 issue of *Astounding Science Fiction.* Smith's story described the Venus Equilateral system, which used three satellites spaced 120 degrees apart to maintain communications between Earth and Venus. As a science-fiction author, Pierce was invited to give a talk on space to the Princeton section of the Institute of Radio Engineers (IRE) in 1954. He chose to talk about satellite communications—a subject that was "in the air" at the time. His talk, later published,[45] posed, possibly for the first time, the choices that would have to be made: passive or active satellites, LEO (low Earth orbit) or GEO (geosynchronous Earth orbit), attitude control, and position control. Pierce provided mathematical analyses suggesting that almost all of the possibilities would work but that some would work better than others. Not least of all, Pierce provided an estimate of the worth of such a system: *$1 billion!*

At the VIIIth Astronautical Congress of the IAF held in Barcelona in October 1957, R. P. Haviland of General Electric presented a paper on communications satellites; he stressed the superiority of equatorial orbits, especially the twenty-four-hour orbit.[46] Haviland proposed a huge manned satellite weighing about 100,000 kilograms and providing 16 television broadcast channels, 2 FM radio broadcast channels, and 60,000 telephone circuits. He felt that solar power would not be feasible and suggested nuclear power as an alternative.

These four authors—Clarke, Burgess, Pierce, and Haviland—carefully described the elements of a communications satellite. Pierce seems to have had the most realistic view of the requirements that a satellite would have to meet to be economically useful. As a senior executive at BTL, he was also in a position to do something about satellite communications. If he could convince his management of the technical utility and economic efficiency of communications satellites, AT&T was in a position to fund development of a satellite communications system and reap the profits.

The Soviet launch of *Sputnik 1* on October 4, 1957, was a surprise that was expected. It was well-known among defense scientists and engineers that the Soviet Union had a powerful ICBM and that the Soviets had promised to launch an Earth satellite as part of the IGY. Even the date had been suggested by a Soviet scientist. The frequencies had been published in amateur radio journals. Still, it was a surprise, especially to the U.S. public and Congress. One reaction was the "militarization" of space. Launching a satellite was now seen as an urgent military and foreign policy priority. The Army was authorized to use the advanced Redstone (*Jupiter-C/Juno I*), to launch a satellite. On January 31, 1958, *Explorer I* became the first U.S. satellite. The Vanguard did not make a successful launch until March 17, 1958.

It has been suggested that the United States was always ahead of the Soviets and that President Dwight Eisenhower saw no need for a "stunt" to prove U.S. superiority. Until both rocket and (reconnaissance) satellite were ready, there was no need to be first. In retrospect, it is difficult to believe that Eisenhower did not think U.S. and world opinion would react in shock to Sputnik. He was personally unimpressed, in part because he knew the United States was ahead of the Soviet Union in space technology and he had been advised that a Soviet launch was highly probable. Perhaps more common was the view of Senator Henry "Scoop" Jackson (D, Wash.), who was quoted in the *New York Times* as describing Sputnik as "a devastating blow to the prestige of the United States as the leader of the scientific and technical world."[47]

Presidents Eisenhower and Truman, for all their differences, viewed government spending as something to be minimized. For this reason, and others, early U.S. space policy, in the words of Walter McDougall, "aimed at sufficiency, not universal superiority."[48] Early in his first term, the conservative Eisenhower had said, "Every gun that is made, every warship launched, every rocket fired signifies, in the final sense, a theft from those who hunger and are not fed, those who are cold and are not clothed."[49]

It is not surprising that Eisenhower would have separated what was to be a "scientific" satellite from the crash military programs to build ICBMs (Atlas and Titan) and IRBMs (Jupiter and Thor). He was apparently anxious both to keep the program open to all scientists and to avoid any delaying interference with the military programs. The satellite decision was made in 1955—a year when committees were advising the president of the perils of Soviet nuclear-tipped ICBMs.

But Eisenhower was right. Except for 1957, when there were no successful U.S. launches, the United States had more successful launches than the Soviet Union from 1958 until 1967. *Almost half of these launches were Air Force reconnaissance satellites*. Except for the Vanguard launches, and later the Scout and Saturn launches, all of the U.S. satellites were launched by converted IRBMs and ICBMs until the Space Shuttle.

By October 1, 1957, all the technologies necessary for satellite communications had been invented. No communications satellites had been launched, but scientific and military payloads had flown in aircraft, missiles, balloons, and sounding rockets. No launch vehicle had put a satellite into orbit, but a wide variety of rockets were immediately available, and higher-performance rockets were being tested. Electronic components were available for the spacecraft payload and the Earth stations. All that remained was to demonstrate these technologies and to compare the different innovations that had been discussed in the technical literature. Industry saw the need for expanded and improved transoceanic telecommunications services and had the funds to develop such services. But the seeds of government intervention had also been sown: the space race was part of the cold war, and AT&T was increasingly perceived as an arrogant monopolist.

3. Post-Sputnik

Industry and the Military Innovate

The U.S. determined to make the Satellite a scientific project and to keep it free from military weaponry to the greatest extent possible. —Dwight D. Eisenhower, October 8, 1957, quoted by Rip Bulkeley in *The Sputniks Crisis and Early United States Space Policy*

The necessary spurs to concrete action [on communications satellites] came with the successful launching of Sputnik I by the USSR on 4 October 1957, and the launching of Explorer I by the United States on [3]1 January 1958. —John R. Pierce, *The Beginnings of Satellite Communications*, 1968

The launch of *Sputnik 1* was followed by a flurry of activity in many areas—including satellite communications. By the end of 1960, Hughes Aircraft Company had built a prototype communications satellite, and AT&T had built prototype subsystems. Hughes was still trying to find a reasonable way of financing its communications satellite system, preferably with an existing communications common carrier as a partner. AT&T was ready to finance its own system. AT&T had already spent tens of millions of dollars of its own funds, and Hughes had spent a few million dollars. The other companies had probably spent somewhere between hundreds of thousands of dollars and a few million. All would be forced to participate in NASA's selection process for a launch opportunity.

The media panic resulting from the surprise launch of *Sputnik 1* was not immediately evident in the public policy process, but it soon began to take effect. The initial impact was on military space programs, but within a few months communications satellites were also being widely discussed. On October 13, 1957, *Aviation Week* reported that a spy satellite project had been under way since 1956 at Lockheed. On November 3 *Sputnik 2* was launched. On November 25 the Air Force awarded a contract for the Sentry reconnaissance satellite to Lockheed. On November 27 both the Thor and the Jupiter IRBMs were ordered into production. By the end of the year, more than a half-dozen Thor IRBMs had been fired. The Thor would later become the workhorse of the space program. The most successful version, the Thor-Able (which, with some modifications, became the Delta), was already designed and would fly in 1958. Like the Redstone variant known *as Jupiter-C/Juno I,* which launched *Explorer I*, Thor-Able was designed to test reentry vehicles. The second and third stages were inherited from Vanguard.[1] In 1958, the Defense Department's Advanced Research Projects Agency (ARPA) was established to manage the space program and to direct high-priority defense technology projects—especially space and missile projects. Roy W. Johnson, of General Electric (GE), was the first director. Leaks and official announcements soon revealed that the Air Force would begin launching reconnaissance satellites by the end of the year. Other early satellite projects seemed slightly Rube Goldberg in design, but more sophisticated projects were planned.

On January 31, 1958, after the Navy-Martin Vanguard's December failure, the Army-JPL *Explorer I*, the first U.S. satellite, was launched by the Army's Jupiter-C rocket. The 15-kilogram satellite had space environment experiments and was placed in a roughly 400-by-3,000-kilometer orbit. On March 17, Saint Patrick's Day, *Vanguard I,* at less than two kilograms (three pounds), was placed in a roughly 800-by-5,000-kilometer orbit. The United States had seven successful launches in 1958. The Soviet Union had only one. Clearly the United States was not behind the Soviet Union in space—except in the weight of the payloads. On May 1 the discovery of the van Allen belts, bands of high-intensity radiation at 1,500 and 15,000 kilometers, was announced. This was the first of many scientific discoveries of the space program. But prestige and science were not the only goals. A congressional staff report covering the hearings on the Space Act, creating NASA, contained a section labeled "The Economic Payoff." The contents of this section seem to emphasize meteorology, but the report also quotes Wernher von Braun as saying: "I am convinced that certain applications of space technology will have very valuable payoffs. Particularly the communications satellite."[2]

While ARPA was temporarily managing the space program, the Eisenhower administration was busy arranging for the old National Advisory Committee on Aeronautics (NACA) to form the core of the new National Acronautics and Space Administration (NASA). At least in part because most of the U.S. space effort was in support of the top-secret reconnaissance satellites, Eisenhower wanted a highly visible civilian space agency to manage scientific launches. NACA had been a research organization; NASA was clearly going to be a mission agency. NACA had performed R&D for industry; NASA would contract with industry to build its launch vehicles and spacecraft. The NACA director, Hugh Dryden, was an obvious candidate for the position of NASA administrator. Unfortunately, he had angered Congress with his low-key, almost denigrating view of the space program and his reluctance to challenge the Soviets in space. In addition, he was a Democrat. James R. Killian Jr. (former president of MIT), the president's science adviser, quickly found T. Keith Glennan and suggested his nomination to Eisenhower. Glennan fit the Eisenhower administration very well: conservative, cautious, internationalist, he understood the place of science and technology in American culture. He was sworn in as the first NASA administrator on August 19, 1958—less than one month after Eisenhower signed the enabling legislation. Dryden would be his deputy.

Glennan had many priorities in his two-year term. The new NASA was very different from the old NACA. New players from the Army and the Navy battled with NACA old-timers for control of the agency while the president and Congress engaged in the same battle from the outside. Glennan would prove that he had the technical, managerial, and political expertise to handle them all.

Glennan was born on September 8, 1905, in Enderlin, South Dakota. Most of his youth was spent in Wisconsin; he went east to attend Yale University. After graduating with a B.S. degree in electrical engineering in 1927, Glennan began work as an engineer in the motion picture sound industry. He installed sound systems in theaters in the United States and Great Britain for Electrical Research Products Inc. (ERPI), a subsidiary of Western Electric, itself the manufacturing subsidiary of AT&T. After a succession of management positions with ERPI in Great Britain and New York, Glennan went to Hollywood.[3]

The war took Glennan from Hollywood to the Navy's Underwater Sound Laboratory in 1941. In 1942 he was appointed director of the laboratory and was brought into contact with the leaders of the Office of Scientific Research & Development (OSRD): Vannevar Bush, from MIT, and James Bryant Conant, from Harvard. After the war, Glennan looked for broader responsibilities; in 1947 he found the ideal job: president of the Case Institute of Technology in Cleveland, Ohio. College presidents have many responsibilities, not the least of which is raising money—especially if they have plans to improve their college. Glennan proved to have a talent for money-raising and for the speaking chores that went along with it. He began to speak on national issues from a conservative position, although he supported both Democratic and Republican candidates. From 1950 to 1952 he was a member of the Atomic Energy Commission during the Truman administration. During subsequent years he served as an adviser to Congress on atomic energy and was appointed to the board of the National Science Foundation. By 1958 he was well-known within the government as an expert on management, technology, and science.[4]

The NASA that Glennan would manage was a strange amalgam. NACA brought with it three research centers: the Langley center in southern Virginia, the Lewis center in Ohio, and the Ames center in California. Langley was the premiere center and had done most of the NACA space work. Lewis emphasized propulsion, including rockets. Ames was the wind-tunnel center. The Naval Research Laboratory (NRL) contributed the Vanguard team, which later formed the core of the NASA Goddard Space Flight Center in Maryland. Within a few months, the Army-CalTech Jet Propulsion Laboratory (JPL) in California also became part of NASA. After a little more than a year, the Army Ballistic Missile Agency (ABMA) in Alabama would also join NASA, eventually becoming the Marshall Space Flight Center. With ABMA came Wernher von Braun, the leading "space cadet." Although aeronautics was not forgotten, it was clear that the new agency would emphasize space technology. The nation expected to see scientific Earth satellites, lunar probes, planetary probes, and manned space flights. Space applications were mentioned, especially weather satellites, but the emphasis was on space science and manned space flight.

Thinking about Communications Satellites

In early March 1958 AT&T's John R. Pierce and Rudolf Kompfner, the independent inventors of the traveling-wave tube (TWT), saw a picture of the shiny 100-foot sphere that William J. O'Sullivan, of the NACA Langley Research Center, was proposing to launch into space for atmospheric research.[5] It reminded Pierce of the 100-foot communications reflector he had envisioned in 1954. The recent invention of the MASER amplifier made reception of reflected communications signals more practical than it had been in 1954. He visited NACA Langley to confirm his understanding of the sphere and by the end of the month was discussing the project with Dryden in Washington. In July Pierce and Kompfner participated in an Air Force–sponsored meeting on communications at Woods Hole, Massachusetts. They were unimpressed with the Air Force plans, which to them seemed unrealistic. While there, Pierce met William H. Pickering, of JPL, who had received his Ph.D. from CalTech the year before Pierce. The three engineers discussed the possibility of launching a sphere such as O'Sullivan's for communications experiments. Pickering volunteered the support of JPL (which eventually resulted in use of the JPL Goldstone station as the West Coast station for Echo). To support this plan,

Kompfner and Pierce wrote a paper,[6] which they presented at an Institute of Radio Engineers (IRE) conference on "Extended Range Communications" at the Lisner Auditorium of George Washington University in Washington, D.C., on October 6–7, 1958.[7]

On October 15, just two weeks after NASA had begun operation, Pierce was invited to serve on an ad hoc ARPA panel on communications satellites, concerned especially with twenty-four-hour communications satellites. Part of the briefing included soldiers using hand-held satellite terminals to communicate on the battlefield. Pierce was upset that requirements and needs were driving military R&D activities with little or no reference to, or apparently knowledge of, the state-of-the-art technology. At this meeting the Department of Defense (DoD) communications satellite program was outlined: first, a spin-stabilized satellite in 1960 (Courier); second, a body-stabilized satellite in 1962; and finally, a twenty-four-hour satellite. Payload frequencies would be both VHF and microwave. Pierce noted that the division of labor—with Army communications and Air Force satellites—was causing some friction. He suspected the Air Force of wanting to take over the whole program. The unreality of the military programs made Pierce only more anxious to begin the Echo program with NASA.[8]

C. C. Cutler, D. E. Alsberg, and Kompfner, all from AT&T and Bell Telephone Laboratories (BTL), visited Space Technology Laboratories (STL), a division of TRW, the following week to discuss the ARPA satellite program. STL was the Air Force systems engineering contractor. With some exceptions, the members of the AT&T-BTL team were unimpressed with the proposed STL-run communications satellite program and with STL personnel. They described some of what they read as "scientific quackery." Cutler concluded a memo to Pierce by saying: "Any hope of BTL taking part in STL's program would appear to be in a position subservient to an irresponsible organization and does not seem desirable to me. There must be a better way to get into the space communication business."[9]

DoD had been the primary space agency before Sputnik, and it remained dominant for some time thereafter. Initial responsibility for space activities was placed in ARPA. In the period between the early 1958 announcement of the intention to form NASA and late 1959, ARPA was responsible for managing all the U.S. space programs and then dividing up responsibility for them among the three services and the newly formed NASA. Satellite communications was assigned to the Army, with the Air Force assuming responsibility for development of communications satellites. This strange division of labor would later cause problems in the Advent program. Since DoD was taking all the active communications satellite programs, only passive (reflector) satellites like Echo were left for NASA.

In November 1958 NASA and ARPA representatives met with the president's science adviser and representatives of the Bureau of the Budget to discuss satellite communications. They agreed that ARPA would concentrate on active satellites and that NASA would concentrate on passive satellites.[10] On November 19 ARPA publicly announced its communications satellite program. A month later, on December 18, the military interest in satellite communications was made more obvious by the launching into orbit of an entire Atlas, with a small communications payload attached. Project SCORE (Signal Communication by Orbiting Repeater Equipment), as this feat was called, involved a 15-kilogram communications payload that could record messages transmitted from Earth and play them back at a later time. Its first message was Christmas greetings to the world from President Dwight Eisenhower. The payload was battery-powered and lasted twelve days.

During late 1958 the House Select Committee on Astronautics and Space Exploration solicited the opinions of "the leading scientists, engineers, industrialists, military officials, and government administrators" regarding the next ten years in space.[11] The results are fascinating, both for the prescience of some and for the wildly inaccurate forecasts, optimistic and pessimistic, of others. Wernher von Braun of V-2 fame, then at ABMA in Huntsville, predicted three accomplishments (in order of occurrence): (1) manned flight to the Moon; (2) twenty-four-hour orbit communications satellites; and (3) weather satellites. Von Braun noted, "Revenues from this worldwide [communications] service should be used for the financial support of future deep-space exploration projects."[12]

The NASA predictions of expected progress covered four areas: scientific, applications, advanced technology, and support facilities. The emphasis in the applications area appears to have been on meteorological satellites, including those in twenty-four-hour "stationary orbits." NASA management defined the NASA program as "aimed at the solution of the scientific and technical problems inherent in each system and the demonstration of practical applications of the systems." Further, the NASA managers explained that how much of their program could be achieved in the coming decade depended "in a large measure on the funding provided."[13]

Many of the scientists stressed weather, communications, and geodesy-cartography as major space applications whose benefit would be seen within the decade. Some of the respondents made strong statements regarding the importance of satellite communications. Herbert F. York wrote, "By the end of the period I expect that satellites will constitute the principal means of intercontinental communications."[14] Arthur C. Clarke stated, "Of all the applications of astronautics during the coming decade, I think the communications satellite the most important."[15] And Louis G. Dunn added, "It is my opinion that the most significant development that will emerge out of the various space programs is the use of satellites for a *worldwide communication system*."[16]

The staff summary of the contributions does not mention satellite communications until more than halfway through but does emphasize the economic returns of this space application Of the fifty-six contributors (NASA had six contributors to one set of predictions), nineteen were from industry. All of these industrial contributors were from the aerospace and chemical (i.e., fuel) industries. None had any special expertise or interest in communications.

Industry, the Military, and Communications Satellite Design

Leonard Jaffe, a ten-year veteran of NACA, was assigned in late 1958 as chief of the Communications Satellite Program at the NASA headquarters in Washington, D.C. After receiving a bachelor's degree in electrical engineering from Ohio State University in 1948, Jaffe had joined NACA as an aeronautical research engineer in 1949. His principal field of endeavor at the NACA Lewis Research Center in Cleveland was electronic instrumentation and data handling. In January 1959 Jaffe formally took on his new responsibilities; in February the Echo program was approved; by April he was testifying before Congress, predicting that synchronous communications satellites would be operational within five to six years.[17] Roy W. Johnson, ARPA director, suggested to Glennan that coordination and information exchange would benefit from a NASA representative on the ARPA Requirements and Technology Panel for Communications Satellites.[18] Glennan appointed Jaffe to the panel.[19]

In early January 1959 Pierce was writing an internal brief soliciting major AT&T support of satellite communications R&D and preparing for meetings with NASA. ARPA and DoD were not interested in his plan to bounce signals off balloons in orbit, but NASA was—especially since NASA was already committed to launching balloons to study atmospheric density. Pierce estimated his funding needs for 1959 as $1 million, with larger amounts needed in succeeding years. On January 9 the AT&T team met with NASA management, including Jaffe, and concluded an agreement: NASA would be responsible for building and launching the Echo balloon; JPL and AT&T would be responsible for building communications facilities[20] and running the Echo experiments.[21]

In February the Air Force launched its first satellite on a Thor-Agena. In March NASA and the Air Force announced details of the all-solid-fuel Scout launch vehicle. By the end of the year the Thor-Able and the Thor-Agena were beginning to replace the original Redstone-Jupiter-Juno and Vanguard launch vehicles. Thor variants dominated U.S. launches until being joined by Atlas variants in 1963. Over 300 of each of these rockets, originally designed as weapons, have been used to launch satellites. Over 100 Scouts have been launched—some as sounding rockets from Wallops Island, rather than as satellite launchers. The Scout was never a communications satellite launcher, but it figured prominently in the early plans of a small group of innovative engineers at Hughes.

In late 1958, possibly due to the success of their ICBM program, the Soviets canceled their advanced intercontinental bomber. In 1959 the Air Force canceled the F-108 long-range interceptor, the purpose of which had been to shoot down the now-canceled Soviet bombers. Hughes was responsible for the fire-control system and the missile for the F-108. Cancellation of the program was a major blow to the company, leading to the layoff of 20 percent of the Hughes employees. Frank Carver, manager of the group that was designing the F-108 fire-control system in El Segundo, California, had seen the coming blow and had earlier asked Harold A. Rosen, a CalTech Ph.D., to explore potential markets for the skills of the Advanced Development Laboratory personnel. In 1959 Donald D. Williams, a physics major from Harvard, suggested to Rosen that a navigation system simpler than ARPA's Transit could be designed using geosynchronous satellites. Rosen felt that GEO (geosynchronous Earth orbit) was more suited to communications satellites. Pierce and Kompfner's article on satellite communications had recently been published, in the March 1959 issue of the *Proceedings of the IRE*. In the article, Pierce and Kompfner had been somewhat negative toward geosynchronous communications satellites because the need for body stabilization, propulsion, large antennas, and high power seemed to require a very large (too heavy for existing launchers), very sophisticated satellite. Rosen felt that a simple, lightweight solution to these problems could be found. For the next several months Rosen and Williams worked on that solution.[22]

Meanwhile, Congress held two days of hearings, on March 3–4, 1959, about "satellites for world communication."[23] Only six organizations were represented, but most of them—NASA, ARPA, AT&T, GE, and ITT—were actively engaged in satellite communications research. AT&T and ITT even expected to make a profit in this endeavor. Perhaps the main message evident in the congressional hearings was that satellite communications was being taken seriously.

Johnson presented ARPA's communication satellite program: (1) SCORE, the Atlas package flown in 1958; (2) Courier, another store-and-forward payload, this one on a real satellite; (3) a polar-orbit real-time repeater; and (4) a twenty-four-hour equatorial orbit (i.e., geosynchronous) real-time repeater. Although the terminology was not used in the open sessions, ARPA was describing the Notus program: Steer, Tackle, and Decree. Steer is especially interesting because it was a UHF communications satellite for the Strategic Air Command (SAC). In the event of war, SAC bombers would head out over the North Pole en route to the Soviet Union. Steer, a system of polar satellites, would provide decision-makers with the ability to communicate with the bombers. Tackle was an intermediate step to Decree, the twenty-four-hour orbit satellite. Tackle would be a similar satellite, in a more easily achievable 10,000-kilometer orbit. The ARPA communications program budget for 1959 was $15 million but would rise to $60 million in 1960 and $100 million per year thereafter. This did not include spending for Centaur, the high-energy upper stage, or for Saturn. The problems of the twenty-four-hour orbit satellite might require the payload capability of this much larger launch vehicle.[24]

Pierce, of AT&T-BTL, made the second presentation. He briefly mentioned the planned NASA balloon reflector experiment and the twenty-four-hour orbit satellite, but most of his presentation and the supplementary statement addressed the technologies required to build satellite communications systems—especially AT&T advances in these technologies. Pierce stressed that in its current work, AT&T was combining the low-noise MASER with a "low-noise" (very directive) horn antenna. Although not elaborated by Pierce, it should be pointed out that the most important characteristic of a communications system is the carrier-to-noise ratio (C/N) at the output of the first stage of the receiver. All later amplification stages will amplify both the signal and the noise—and add some more noise. The AT&T engineers seem to have worried most about minimizing noise instead of maximizing signal, presumably on the grounds that maximizing the satellite-to-ground signal implies complexity on the satellite and that minimizing the noise implies complexity on the ground. AT&T also investigated different modulation techniques (SSB, FM, PCM), solar batteries (i.e., aggregations of solar cells), thermal control, high-frequency transistors, thermal control, orbits, and long-life microwave tubes. It was also installing a 60-foot microwave antenna and powerful transmitter. Working on satellite communications were a few dozen engineers, scientists, and technicians, of whom about one-quarter were full-time.[25]

In a supplementary statement, Pierce outlined AT&T's contributions to communications electronics. AT&T had pioneered electronics, originally to develop transcontinental telephony at the turn of the century and later to develop intercontinental telephony in the 1920s. AT&T's wartime work on microwave radar was based on prewar experimentation and led to the TD-2 microwave system for long-distance telephony—and network television—in 1947–51. AT&T also built tropo-scatter systems for the DEW line and the "White Alice" communications system in Alaska. BTL had conducted numerous studies of propagation at many frequencies. Pierce did not mention BTL's development of the transistor or his own co-invention of the TWT, both necessary to satellite communications. In any case, his message was obvious: AT&T could do the job and in fact was already started.[26]

Haviland spoke for GE. He began by recounting his experiences in the 1945–47 effort of the U.S. Navy (Bureau of Aeronautics) to get into the space business and his subsequent move to GE, where he became the in-house "satellite expert." Haviland was convinced that the communications satellite was the "most important single application of the immediate, close-to-the-earth space vehicle." Haviland described two potential communications satellites: one for the future, and one for now. His future satellite was a huge, manned television broadcast space station with toroidal (presumably rotating to produce artificial gravity) living quarters and massive antennas. His budget was expected to be less than $2 billion. Haviland also described a second, less ambitious project, suggested by his GE colleagues. This was a 2,000-kilometer-orbit satellite with Sun-tracking "solar absorbers" and an Earth-tracking antenna. Sixteen of these satellites would cover Earth. Each satellite had four transmitters, each capable of 200 simultaneous teletype messages. The whole would cost from $100 million to $150 million to establish and perhaps $50 million per year to operate. Assuming that 50 teletype channels are equivalent to one voice channel and that half the satellites are in useful positions at any given time, the system provided a little over 100 voice channels for slightly more than $100 million. By comparison, the TAT-1 telephone cable initially provided 24 channels (later 36 channels) for about $50 million. Haviland assumed that his 250-kilogram satellite would cost only $2.5 million to build and $2.5 million to launch on an "Atlas-type" booster. Based on later data, his numbers are probably low by a factor of four. Whether or not the plan was cost-effective, GE was also prepared to enter the communications satellite enterprise.[27]

Henri G. Busignies, president of International Telephone and Telegraph Laboratories (ITT), presented a description of ITT itself and discussed the company's ongoing research in satellite communications. ITT was primarily involved in international communications operations and electronics manufacturing. The company had surveyed the possibilities for satellite communications for some time and had recently been concentrating its research on twenty-four-hour satellite systems. ITT was comfortable with the communications components for both satellite and ground station but had joined with Curtis-Wright and Aerojet for assistance with the satellite vehicle itself. Busignies expressed particular concern about attitude- and orbit-control problems. ITT cost estimates ranged from $100 million to $200 million for twenty years of operation for 500 voice channels on a 350-kilogram satellite. Busignies pointed out that these cost estimates bracketed the cost of submarine telephone cables. Thus ITT was also interested in communications satellites, but it seemed to be looking for a partner in this venture—especially a partner who understood satellites.[28]

Edgar M. Cortright, accompanied by Newell Sanders and Leonard Jaffe, made the NASA presentation. Cortright began with a discussion of the growth in transatlantic telephone messages. He noted that the current cable capacity would be exceeded in 1962 and that advanced cable capacity would be exceeded in 1965 due to the high growth rates. He then briefly described the passive satellite program then under way with AT&T and a 400-1,500-kilogram, body-stabilized active repeater in a twenty-four-hour orbit. Most interestingly, he described the technology-development areas for the two systems. The passive system technologies were all ground-station technologies. The active system technologies were all spacecraft technologies. NASA was aware of satellite communications, but its commitment was not as evident as that of ARPA, AT&T, GE, and ITT.[29]

The statements and questions of the members of the House Committee on Science and Astronautics ranged from the profound to the puerile. It seems that these congressmen, with limited technical understanding, were nonetheless excited about the potential benefits of a commercial communications satellite system. They had many questions not addressed by the speakers. One concerned the issue of ownership and operation. Should this commercial system be owned and operated by the government? Should the government put up the initial system? The committee members repeatedly asked for costs, profitability, and timetables. Were the companies actually spending their own money now? The congressmen were also concerned about jamming or other use by the Soviets or others. Two important points were made by ITT's Busignies. He explained (again) that all the communications electronics had been developed for both ground and spacecraft—what was missing was a satellite vehicle (bus) to carry the communications payload. Busignies also argued that active satellites were ultimately cheaper than passive—and that twenty-four-hour satellites were the best active satellites. None of the other speakers really challenged these comments. Pierce almost got into an argument because of his conservative statements: engineers, even "executive" engineers, are not used to posturing. Not all of the industry representatives were comfortable dealing with Congress, but it is clear that at the end of the hearings, the House Committee on Science and Astronautics was fully aware that industry was spending its own money on communications satellite research and that commercial systems were not far off.

The staff report based on the hearings seems to have taken a position slightly different from that of the transcript of the hearings themselves. If this position truly represented the feelings of the committee members, it is quite significant. After a general introduction to the problems of communications in the modern world, the report addressed the military requirements, seen to be best met by the twenty-four-hour equatorial repeater. The major problem identified was the need for a heavy-lift launch vehicle. This problem was forecasted to be solved in the near future by the Atlas-Centaur—and ultimately by the Saturn. In discussing civilian requirements, the report stressed the consensus opinion that valuable commercial operations were possible in the immediate future but that the probable high costs had created a mood of caution—albeit optimistic caution. More important was the conclusion that "work leading toward commercial application should start as a government enterprise." The report concluded with a review of the technologies described by AT&T and the various cost estimates. The idea had already begun to germinate within Congress that this was an enterprise that should be directed, at least in its early stages, by the government.[30]

But AT&T, as the world's largest communications company, clearly felt that it should lead this endeavor. ARPA, at least in the opinion of the AT&T-BTL engineers, was too ambitious—and perhaps insufficiently knowledgeable. AT&T, JPL, and NASA Langley were proceeding with a very simple passive experiment that, if successful, would lead to more complex satellites. AT&T clearly saw itself as the leader on the passive program. The most important ground facilities would be at AT&T-BTL in New Jersey and JPL in California. The rest of NASA would simply build and launch the balloon. AT&T engineers treated JPL as a knowledgeable junior partner but the rest of NASA as not very competent. Their worst private comments were reserved for ARPA engineers and managers, whom they saw as unrealistically grandiose in approach.[31]

In the June 22, 1959, issue of *Aviation Week*, an article on civilian communications satellites described the efforts of AT&T, GE, ITT, RCA, Space Electronics, and Westinghouse, as well as ARPA efforts. The AT&T, ITT, and GE material is essentially that presented in March to Congress. The efforts of Space Electronics are highlighted in the article, which seems to have derived from interviews with the Space Electronics president: Dr. James C. Fletcher, a future NASA administrator. The cost of establishing a 250-voice channel, single television channel, twenty-four-hour equatorial system, according to Fletcher, would be between $25 million and $40 million. The proposed system would weigh approximately 500 pounds and be launched by the NASA Vega launch vehicle. This launch vehicle, subsequently canceled due to the availability of the Agena upper stage, would have consisted of an Atlas first stage and the Vanguard first stage configured as a second stage. The third stage would be a solid. Fletcher also proposed using ground antennas similar to the JPL Goldstone 85-foot parabolic reflector, dipoles for spacecraft attitude sensing, and small control thrusters for attitude control.[32]

In August DoD announced that two contracts had been awarded for work on the Notus program. GE received a $5.5 million contract to start development of the spacecraft. Bendix Aviation received an $8.5 million contract to start development of the communications subsystem. The first proto-operational system had been initiated.[33]

The Early Designs

Rosen and Williams had been busy at Hughes during the first half of 1959, and they had been joined by Tom Hudspeth, an experienced electrical engineer. By the summer they had designed, at least conceptually, a lightweight communications satellite and were ready to make presentations to upper management at Hughes. On September 17 A. S. Jerrems informed Frank Carver, Rosen's boss at the Radar Laboratory, that a presentation to Dr. Allen Puckett, Hughes executive vice-president and assistant general manager, would take place on September 23 at 11:00 A.M.—later rescheduled for 2:30 P.M. on September 25.[34] One result of all the meetings was a formal evaluation by S. G. Lutz. On October 1 Lutz reported his results to A.V. Haeff, vice-president, research.[35] His major recommendation was that Hughes should seriously consider a commercial venture in satellite communications. Lutz felt that Rosen and Williams were overly optimistic, but he also thought that their lightweight design of a Scout-launched twenty-four-hour stationary orbit communications satellite could be a prestige-generating accomplishment for the *company* that designed and launched it.

Lutz had two major complaints about the Rosen-Williams proposal: difficulties in orbit control, and overly optimistic cost estimates. The orbit-control complaint was rather broad—Lutz was concerned both with Scout performance and with spacecraft on-orbit control. The satellite weight was somewhat optimistic. Lutz pointed out the lack of weight estimates for solar cells and chemical batteries (for eclipse operation). The Scout was an all-solid fuel rocket with the high accelerations implied by that choice of propulsion. Additional structural weight might have been required to withstand the launch forces. The in-orbit propulsion system consisted of a .22-caliber "gun" that fired as it spun past the appropriate direction in space. Lutz did not think this would work. In addition, Rosen and Williams estimated total costs at approximately $5 million. Lutz compared this with a recent proposal to Hughes from Fletcher, of Space Electronics, to cooperate in a $45 million communications satellite venture and with the recent award of an $8.5 million contract for the communications package alone on the "Air Force" satellite.

In addition to technical and cost issues, Lutz saw one further, potentially overwhelming business issue: "Though a 4.5 mc [MHz] satellite link would have the fantastic potential value of about a billion dollars on this basis [$200,000/kc], its actual investment value would depend on its earnings from the traffic it could be sold for. As Shakespeare observed, 'here is the rub!' HAC [Hughes Aircraft Company] is not a Communications 'Common-Carrier,' is not apt to become one and might have difficulty in negotiating with one. Even the Bell System could not today supply 4.5 mc of trans-Atlantic phone traffic, nor would there be more than intermittent use for television."[36] Lutz recommended a thorough task-force study concentrating on the "business" aspects of satellite communications. He further recommended talks with General Telephone, the largest independent telephone company.

The Hughes Task Force on Commercial Satellite Communication was formed shortly after the Lutz memo. Its first meeting was held on October 12, 1959. The meeting determined that the most important aspects of the communications satellite program were the economic ones rather than the technical ones. The AT&T monopoly would be difficult, but not necessarily impossible, to circumvent. The value of a satellite was not necessarily determined by its ability to carry telephone traffic; it could generate corporate prestige, stake a claim on a section of the stationary orbit, transmit television or provide other wide-bandwidth service, and possibly provide military communications. The Air Force and Army systems were seen as noncompetitive, but what about AT&T and RCA? What were their plans? Should government support be sought?[37]

On October 22 the task force reported its findings: "It is the unanimous opinion of the Task Force working members[38] that the satellite communication system proposed by Dr. H. A. Rosen is technically feasible, is possible of realization within close to the estimated price and schedule, has great potential economic attractiveness and should not encounter too serious legal or political obstacles."[39] The unanimous report went on to detail other technical, economic, and political issues. Lutz was unsure what communications traffic the satellite could capture, but he was sure that with a 200:1 cost advantage, it would not go unused. This cost advantage, according to Lutz, was not due to any technical breakthrough but rather to the "Hughes brand of System Engineering." Everyone else (NASA, RCA, Space Electronics, the Army, and AT&T) had seen the problems of GEO as solvable only by large complex spacecraft. Only Hughes had attempted (successfully) to design a cheap, lightweight spacecraft. The task force had made some suggestions for improvements; these included better spin-stability, a stiffer structure, and provision for access to the transponder by multiple Earth stations (multiple-access).

The task force had five recommendations:

1. Start now! Prestige will go to the first satellite.
2. Fund the traveling-wave tube separately as a commercial product.
3. Fund the remainder of the project at a level of about $850,000.
4. Explore cooperation with General Telephone.
5. Form a project team to carry out the program.

A separate memo from J. H. Striebel reported the results of contacts with General Telephone of California, KABC, Pacific Telephone (Bell), and the Federal Communications Commission (FCC). The talks were inconclusive, but requests for data were accepted, and most of the companies reported a willingness to use any high-quality, low-cost facility made available to them. Striebel recommended first getting clearance from the government agency that had "cognizance over communications satellites." This should be followed by discussions with communications companies regarding cooperation. If these failed, Striebel advised, Hughes should propose the system to the government as an interim "task [D]ecree" (Notus) program.[40]

On October 26 Haeff arranged for a presentation to L. A. (Pat) Hyland, the Hughes general manager. Hyland subsequently made four recommendations:[41]

1. Determine patentability.
2. Coordinate with NASA in a manner similar to "AEC projects in which the government finances the development of a type or class of reactor but allows industry to proceed with the commercial development and to retain commercial rights in the patents that it generates."
3. Do not engage in dialogue with potential "partners" until after steps 1 and 2 are completed.
4. The traveling-wave tube should be developed, but only if it is comparable in value to other research projects.

By the end of the month, a "non-proprietary" description[42] of the Rosen satellite system had been prepared for use in discussions with industry, along with a more complete description[43] for internal use and possible government discussions. Three elements of this design were seen as patentable novelties: the despin method, the orbit-correction method, and the SSB-to-FM conversion.[44] Interestingly, none of these elements (although orbit correction was part of it) represented the "Williams patent" novelties (e.g., the use of a sun sensor to trigger thruster pulses), which were disputed for years to come.[45]

On November 5, 1959, Williams traveled to NASA Headquarters, where he spoke with Homer J. Stewart. Although most Hughes senior executives seem not to have been too upset at the thought of NASA taking over their patent rights, Williams prefaced his discussions with a statement that Hughes did not want to lose its proprietary rights by talking to NASA. Stewart assured him that this would not happen. Williams emphasized the Hughes interest in proceeding with its program as a commercial venture. Stewart supported the idea of a Hughes-funded commercial communications satellite program as compatible with the traditional U.S. approach to telecommunications. Stewart warned

Williams, however, that there was a congressional faction that would oppose privately owned commercial communications satellites.[46]

Back at Hughes, Executive Vice-President Puckett initiated a thirty-day study in November under Dr. R. K. Roney. General Manager Hyland decided that if further analysis proved the program feasible, Hughes would attempt to obtain a contract from NASA for development and launch of the satellite. Patent filings would precede any contract with NASA. Hyland decided that Hughes *would not* spend its own money building a satellite before contracting with NASA.[47] Nor would Hughes attempt to become a common carrier. Nor would Hughes fight NASA over patent rights. But these last three statements of policy were all eventually overturned. Hughes would actually use its own money to build a prototype in 1960 and early 1961. Hughes would become a common carrier in the 1980s. And Hughes would fight NASA for thirty years over patent rights.[48]

A considerable amount of analysis and redesign took place in the closing months of 1959. The design incorporated in the October 1959 "Commercial Communications Satellite" was much less sophisticated than the January 1960 design.[49] The elements that would compose the "Williams Patent" were all evident by January 1960. By the end of 1959 all of the elements of the satellite appeared to be feasible and practical, with one exception: the difficulties involved in using Jarvis Island in the Pacific equatorial region as a base from which to launch the satellite on a Scout rocket. This method was more expensive than estimated and probably more difficult and riskier.[50]

During most of 1959 AT&T concentrated on building the antennas, transmitters, and receivers required for the Echo program. By November, successful experiments had been conducted by moon bounce between the AT&T-BTL facilities at Crawford Hill and the JPL facilities at Goldstone.[51] AT&T and several other companies interested in satellite communications (notably ITT, but also Hughes) looked on the ground facilities as the most important component of a communications satellite system. In general, from this early period to the present day, more money has been spent on Earth stations than on communications satellites themselves. Most of the noise in a satellite communications ground system comes from the first stage of amplification. Since both signal *and* noise are amplified by later stages, it is important that this stage be low-noise. AT&T had pioneered the development of MASERs after they were invented by Charles Townes at Columbia University in 1954. MASERs provided very low noise amplification. A secondary contributor to system noise is "sky noise" received by the antenna. When the antenna is pointed more or less straight up, this noise is minimized. Most antennas have sidelobes—they also receive in directions other than the main lobe (beam). The worst of these can be the backlobe, which points at the "hot," and therefore "noisy," Earth. The AT&T solution to this problem was the horn antenna, which had minimal sidelobes and no back-lobe. Another problem involved tracking the satellite—keeping the antennas pointed in the right direction at an object moving at eight kilometers per second.[52]

Active satellite work had not been neglected. In an August 1959 memorandum, Leroy C. Tillotson described a satellite design quite similar to what later turned out to be Telstar. A separate group had developed a design for a satellite TWT, described in another BTL memorandum. By the end of the year, studies of spacecraft power systems (solar cells, Ni-Cd batteries, DC-DC converters), structures, space environment, thermal control, and attitude control had also been completed. Perhaps more important, a commitment to developing active satellites was building. In the words of Pierce: "By the end of 1959 our thoughts were directed toward a simple, low-altitude *active* satellite as the next step."[53]

On September 23, 1959, the Air Force was assigned primacy in military space activities. All long-range missiles, except for Polaris, would be under Air Force control. The ARPA Midas and Samos projects were transferred to the Air Force, Transit was transferred to the Navy, and Notus was transferred to the Army. Within a month the head of ARPA, Roy Johnson, resigned. The head of ABMA, General John B. Medaris, was furious over the transfer of control to the Air Force. As Wernher von Braun's boss, Medaris may have had the most capable space group in the nation. The Army had been the leader in space, from the postwar V-2 launches through the various JPL rockets, the Redstone missile, the first U.S. satellite in space, and the source of much of the studies and actual work on satellite communications. Both the 1958 Score and 1960 Courier were Army projects. In his 1960 memoir, *Countdown for Decision,* Medaris excoriated Herbert York (deputy secretary of defense for research and development), all civil service civilians in DoD, the Air Force (which he argued should be recombined with the Army), and all businessmen—especially those in the aerospace industry. Medaris also resigned.[54]

In December the Notus program was reoriented. Both Courier and Steer were canceled. All resources would be dedicated to the twenty-four-hour satellite program.[55] A few months later, U.S. Air Force General Bernard A. Schriever took issue with York, as had General Medaris. Schriever was specifically upset by the speed with which York and ARPA were developing the Midas, Samos, and Notus satellites.[56]

By the end of 1959, Hughes had designed the basic satellite that would become Syncom. AT&T and NASA were finishing up their preparations for Echo. AT&T was also proceeding with an "active" satellite design, but most of its effort was going into Echo. DoD had made the decision to put all of its efforts into one component of the Notus program. The geosynchronous Decree, renamed Advent, would be the DoD entrant in the communications satellite race.

Table 3.1
Syncom Mass by Subsystem, 1959 and 1960

Subsystem	Mass (pounds) October 1959	Mass (pounds) January 1960
Electronics	4.0	4.2
Power	5.0	5.3
Antenna	0.6	0.5
Propulsion and Attitude Control	2.1	8.4
Harness	1.0	1.2
Structure	9.5	3.4
Total	22.2	23.0

In January 1960 Roney, of Hughes, reported interim results of the communications satellite analyses to Executive Vice-President Puckett. In general there were no technical hurdles, and the recent redesign of the spin-orientation techniques had made the spacecraft simpler and more robust. Roney made no mention in his memo of the communications market but rather suggested that this program was oriented toward a NASA contract. The only unresolved issues, in Roney's mind, were the Jarvis Island launch site and the tracking network. Both of these problems might be solved by NASA. Also in January, details of the redesigned spacecraft were circulated. The principal changes were the addition of sun sensors, a new structure, and the replacement of the .22-caliber "gun" with a nitrogen gas propulsion system. A comparison of the October 1959 and January 1960 weights is shown in Table 3.1. This was essentially the design that would be launched three years later. The weight would grow substantially, but the general design was converging to what would eventually be called Syncom.[57]

During February 1960 a variety of contacts were made with NASA Langley and NASA Headquarters to discuss the Hughes satellite and the status of the Scout launch vehicle. The meetings were cordial, but not rewarding. Scout was still in development. First launch would be later in the year, from the Wallops Flight Facility (WFF) in Maryland. Because of the "no funding" ruling by Hyland on December 1, 1959, and the somewhat disappointing lack of interest at NASA, Rosen, Williams, and Hudspeth considered striking out on their own. Rosen approached his friend Tom Phillips at Raytheon, who invited the Rosen-Williams-Hudspeth team to join him in Massachusetts, where he would support their project. The team requested that the Hughes interest in their patents be released to them if Hughes did not pursue communications satellites commercially. By the time this request reached upper management, however, Hyland had decided to fund the project. By March General Manager Hyland and

Executive Vice-President Puckett were firmly behind the communications satellite project. Puckett suggested that Hyland approach acquaintances at General Telephone and CBS.[58]

In April 1960 AT&T began to prepare a follow-on passive experiment after Echo. Huge 3,600-square-foot horn antennas would be built on either side of the Atlantic. Forty-kilowatt transmitters would beam television signals up to a duplicate of Echo in a higher orbit—2,000 miles. Unfortunately, the studies showed that passive satellite television transmission would be of marginal signal quality. In a May 13, 1960, letter to Jaffe at NASA Headquarters, Kompfner described the current AT&T-BTL research program as shifting to *active* satellites. Kompfner reviewed the active satellite component-subsystem studies that had been under way since late 1959.[59]

During April Hughes was still ramping up its communication satellite effort. An important part of this effort was finding a customer to help fund the program. On April 1, 1960, Puckett met with York, defense director of R&D, and Richard Morse, Army director of R&D. Technical materials were to be delivered to Dr. E. G. Witting by April 15. Through some misunderstanding, the delivery of these materials, by J. W. Ludwig and F. D. Vieth, became a formal presentation. Not prepared for this, the Hughes representatives were unable to answer questions but took note of all the issues raised and promised to answer them at a later meeting. Several criticisms of the design were made. Some of these reflected DoD ignorance of the Hughes approach; others may have reflected Hughes ignorance of progress made elsewhere. In addition, the Army representatives made it clear that Hughes was a latecomer. Later in May, answers, analyses, and other additional pieces of information were forwarded to the Army. On a trip in May to Fort Monmouth to discuss the Hughes satellite, Hughes engineers were made aware of the soon-to-be-released RFP for the Advent twenty-four-hour satellite. This was, unrealistically, seen as an opportunity. Ultimately, the winners of the study contracts in the previous year, GE and Bendix, received the Advent contracts.[60]

Government Control

The 1959 World Administrative Radio Conference (WARC) in Geneva reserved some bands for space communications, but these were primarily for experimental satellite telemetry and command. The first international call for satellite communications frequencies was made by the United Nations Committee on the Peaceful Uses of Outer Space (COPUOS) in 1960. AT&T, the main promoter of submarine telephone cables, was beginning to make arrangements with its foreign cable partners to test a medium-altitude experimental communications satellite. Hughes was marketing its Syncom satellite to anyone who would listen. DoD, after the Score and Courier experiments, was building Advent, a large geosynchronous satellite. NASA had a representative, Jaffe, on the ARPA Requirements and Technological Panel for Communications Satellites, but other than the Echo project, which now looked trivial, NASA was not a player in the communications satellite arena. Congress had queried Glennan and Dryden on NASA's plans during the fiscal year 1961 (FY61) budget hearings in March 1960. Dryden had replied that policy would be for Congress to decide—especially in regard to property rights.[61]

The FCC, primarily in response to the 1959 WARC actions, began soliciting opinions in May 1960 (Docket 11866) on the frequencies to be allocated to space communication. These were shared frequencies (they were also used for terrestrial microwave relays), and the FCC was preparing for the 1963 Extraordinary Administrative Radio Conference (EARC), at which these issues would be addressed. AT&T had filed a petition for reconsideration of Docket 11866 (known as "over 890") in August 1959. On July 11, 1960, AT&T responded to the reopening of the frequency question by outlining its plans for a global communications system consisting of fifty low-altitude satellites in random polar orbits. Pierce described the Echo program, the difficulties of establishing a twenty-four-hour "stationary" satellite, and the imminent availability of reliable space-borne electronics. The satellite system, described by Charles M. Mapes, consisted of unoriented active repeaters in 3,000-mile orbits. The frequency band was assumed to be 500 MHz wide—capable of handling 600 voice circuits or 2 television channels. The total cost would be about $170 million, which would be funded completely by AT&T and its international partners. J. B. Fisk, president of BTL, provided the conclusion: a plea that the frequencies "above 890" MHz not be allocated to private microwave systems but be preserved for satellite communications. The FCC rejected the AT&T arguments and let the previous ruling stand.[62]

On May 24–26, 1960, the Armed Forces Communications Electronics Association (AFCEA) held its annual convention in Washington, D.C. The highlight of the second day was a panel discussion on "space communications" by three AT&T-BTL speakers: Pierce, "Problems of Satellite Communication"; W. J. Jakes, "Bell Laboratories Part in the *Echo* Experiment"; and L. C. Tillotson, "Active Satellite Repeaters." The talks covered the history of long-distance telephone communications and the specific advantages and problems of satellite communications. The AT&T-BTL speakers seemed to have high hopes for twenty-four-hour "stationary" communications satellites, but they assumed that these satellites were likely only in the future, due to the difficulties of controlling attitude about all three axes and the lack of a heavy booster. Lutz, of Hughes, noted that AT&T had not considered the advantage of single-axis orientation using spin stabilization, which would allow a substantial amount of antenna gain.[63]

After the conference, Lutz spoke with Jaffe about the reopening of FCC Docket 11866. One of the issues was the coexistence of satellite and land microwave communications using the same frequencies—a topic partially covered by Pierce at the conference. It was clear that low satellite power would not interfere with ground microwave systems, but Lutz pointed out that very minimal restrictions on ground microwave systems could avoid their pointing at the geosynchronous or "stationary" arc. Jaffe was apparently unconvinced that "stationary" orbit was the ultimate location for communications satellites. He was concerned that the 0.13-second delay between Earth and satellite due to the speed of light would be intolerable in a two-way conversation in which 0.5 seconds would elapse between the time a speaker finished a sentence and then heard the response. Lutz found Jaffe to be cautious and noncommittal, as if he were afraid of taking a position. Jaffe mentioned studies being performed for NASA by RAND and Stanford Research on the economics of satellite communications. The two men agreed that AT&T's position on FCC Docket 11866 seemed to be political rather than technical.[64]

Hughes had been spending about $100,000 per month on the "Commercial Communications Satellite" program since April. By the beginning of July, about $300,000 had been expended. Estimates for the "demonstration of a commercial system" were beginning to climb to about $15 million. The satellites were quite inexpensive, but Earth stations and Scout launch vehicles had to be purchased, and a launch facility had to be prepared on Jarvis Island. Clearly, the Hughes chosen strategy was to get NASA to fund a demonstration of its system, which it would then market commercially. This had been Hyland's suggestion at the October 26, 1959, meeting. Other tactics had been and were being tried, but this one seemed most promising. Even without NASA funding, Hughes seemed to have been willing to spend substantial sums of its own money. Hughes was convinced that its satellite would eventually succeed in the marketplace and was willing to take the risks necessary to stay in the game.[65]

The advantages of twenty-four-hour stationary orbits had been touted for some time by Hughes, which now began to compare the advantages of its system with those of the AT&T-BTL system. The traditional advantages of the stationary satellite were fewer satellites for global coverage (three rather than fifty) and simpler, cheaper Earth stations (no "tracking" required, $100,000 antennas versus $500,000 antennas). Both Hughes and AT&T thought their satellites would cost about $1 million. Other organizations estimated the cost of satellites, especially stationary satellites, at as much as $25–50 million. In the

June proposal to NASA, the reliability possible with the benign radiation environment at synchronous altitudes was contrasted with the damaging radiation environment at the low altitudes envisioned for the AT&T-BTL satellite. The June 1960 revision of the October 1959 satellite design document reflected the growing sophistication of the Hughes design and the use of a Thor-Delta fired from Florida rather than the Scout fired from Jarvis Island. In addition, the title had been changed from "Commercial Communications Satellite" to "Synchronous Communication Satellite."[66]

In May Williams published his analysis of the orbital dynamics and attitude dynamics of the Hughes synchronous satellite. The document provided the analytical basis for the "Williams Patent." Copies of this document were sent to NASA Langley, the Naval Ordnance Test Station, and other potential customers. Williams also studied orbit determination based on angle and range rate data for Ford Aeronutronic.[67] A discussion of this study with Samuel Herrick and Lou Walters of Aeronutronic provided Hughes with information on orbit-determination methods with which Williams and his staff were unfamiliar.[68]

For the rest of 1960, Hughes continued to make presentations to likely customers. These customers included NASA, DoD, GTE (General Telephone & Electronics), ITT, and even AT&T. The NASA presentations were the most numerous: to Glennan (August 16), Jaffe (July 12, September 1), Siegfried Reiger, of RAND (September 26), and Fred Morris, of Stanford Research (November 4). DoD presentations were not far behind: to Harold Braham, of Aerospace Corporation (October 26), General Schriever (November 3), and Schriever's Ballistic Missiles Division staff (November 8). Meetings were held with GTE in August and September. Hughes personnel also made presentations to AT&T and ITT in November. Hughes seemed to put most of its effort into obtaining a NASA contract but was also very interested in finding a common carrier with whom it could enter into the satellite communications business.[69]

Echo 1 was launched into a 1,000-mile circular orbit on August 12, 1960. During the first orbit of the 100-foot sphere, a voice recording of President Eisenhower was transmitted from JPL's Goldstone, California, Earth station to AT&T's Holmdel, New Jersey, Earth station. Later experiments included telephone, teletype, and facsimile transmissions. On August 23 the first transatlantic voice transmission was executed from Holmdel to Jodrell Bank, England. In spite of the success of *Echo 1,* it was clear that active, rather than passive, satellites were the technology to develop.[70]

AT&T began serious work on an active satellite during the summer of 1960. On September 1 a meeting was held at 195 Broadway, Corporate Headquarters, in New York City, to discuss the AT&T program and its relationship with NASA. A report at the meeting stated that on August 11, senior AT&T executives had briefed NASA on AT&T discussions with the communications administrations of Great Britain, France, and Germany, highlighting the AT&T view that satellite communications in the United States would be a private enterprise initiative. In addition, on August 30 G. L. Best, vice-president at AT&T, had met with representatives of the State Department to tell them of AT&T's plans and to report its discussions with NASA. The remainder of the meeting was a discussion of the direction that AT&T's efforts should take. The consensus was that AT&T should pay all costs of the trial of a 6,000-mile-orbit satellite with a capability of 600 telephone channels. Foreign partners should pay for their own Earth stations but should not be asked to defray the costs of the satellite. Service would not be available until the mid-to-late 1960s, so submarine cable plans should proceed without change. Initially, a reimbursable NASA launch would be the simplest way into orbit. Brief mention was made of the Hughes 20-pound repeater and the ongoing Advent competition.[71]

On September 15, 1960, Best hand-delivered a letter to NASA Administrator Glennan. The letter outlined AT&T's plans for communications satellites. Best described AT&T's ongoing work on an active satellite and Earth stations and noted its hope that an experimental transatlantic satellite could be launched into a 2,200-mile orbit within eighteen to twenty months. The communications package would be a 5-MHz repeater. Best said that AT&T would assume all the costs in this trial except for the foreign Earth stations. He added, "[We] would hope that the National Aeronautics and Space Administration would be willing to launch these trial satellites for us, at our expense, if this proved to be the most practicable arrangement." Glennan replied, on September 28, that "issues of national policy" raised by the AT&T proposal were still being studied and that he was thus prevented from responding. On October 12 Glennan made a speech in which he stated that U.S. policy would be to allow private industry to develop satellite communications. On October 20 E. I. Green, executive vice-president at AT&T, forwarded to Glennan a description of the technical features of the AT&T satellite. On October 21 AT&T filed with the FCC a request for permission to launch and operate a satellite communications system.[72]

Meanwhile, the DoD communications satellite program was beginning to look a little shaky. The Advent program had organizational and management problems that may have been worse than its technical problems. The Air Force felt that space was just an extension of the Air Force mandate, with no need for the involvement of the other services. The Army, still smarting from its loss of ballistic missiles and space research (ABMA), felt that it certainly had more communications expertise than the Air Force. ARPA was caught in the middle. John Rubel, an ex-Hughes employee and deputy director of ARPA, made the decision to assign Advent to the U.S. Army Signal Corps effective September 15,

1960—thus implementing a memo from the secretary of defense on September 18, 1959. The failure of the first Courier satellite due to a Thor first-stage malfunction on August 18 did not improve matters. The Centaur upper-stage for Atlas was also underperforming. The Atlas-Centaur capability for synchronous orbit was now 750 pounds—somewhat under the nominal 1,000-pound weight of Advent. *Courier Ib* was finally launched successfully on October 4, 1960. It operated very successfully until October 22, when the transmitters failed. Initially funded at $8 million, the program had eventually cost $20 million. Advent estimates were just under $200 million. The Army rejected, apparently without understanding, an unsolicited proposal from Hughes.[73]

During 1960, AT&T and Hughes had begun to market their programs. Both seemed to be very close to an operational system. Many elements of the government seemed to feel that this was something for which no policy existed. Who in the government was in charge of satellite communications? Aside from the Echo project and some comments on RAND studies of the "economic, political and social implications of space research,"[74] the first reference in NASA Administrator Glennan's diaries to communications satellites is an entry for Wednesday, July 27, 1960, in which Glennan refers to a meeting with George Kistiakowsky, the president's science adviser, to talk about communications satellites. The concern was that public policy had not been developed in this area and that the "pressures generated by AT&T and the military, as well as other industrial suppliers," were "building up quite a fire."[75] The next day Glennan assigned Robert G. Nunn, NASA assistant general counsel, the task of preparing the outline of a paper—to be delivered at a cabinet meeting—asking that NASA be given the responsibility for preparing administration policy on communications satellites. A week later Nunn and John Johnson, NASA general counsel, met with Glennan to discuss "the communications satellite problems." NASA had concentrated on passive satellites, as specified in the 1959 division of labor with DoD, and had assumed that DoD would be responsible for active satellites. On this same day, August 2, 1960, Glennan had lunch with William O. Baker, head of BTL, and on the afternoon of August 3 listened to a presentation by NASA personnel on the active communications satellite. On August 8 Glennan spoke with Jaffe and got him to admit that he had overstated his claims for the civilian communications satellite program. On August 11 Glennan met with an AT&T delegation that had been exploring a joint communications satellite program with partners in England, France, and West Germany. A few days later he was informed by the White House that Hughes was interested in NASA support for its communications satellite project. On August 16 Glennan listened to presentations by both Hughes and AT&T on active communications satellites. In the space of a few days NASA, at least as evidenced by Glennan's diary entries, had gone from a position of almost no previous interest in communications satellites to one of very high interest—even to "owning the problem."[76]

After consultation with Maurice H. Stans, director of the Bureau of the Budget (BoB),[77] and James H. Douglas Jr., deputy secretary of defense, Glennan developed a NASA-DoD agreement that allowed NASA to proceed with an active communications satellite program. Discussions with NASA General Counsel John Johnson made it clear that there were many policy problems to be solved. During this period *Courier 1a,* the first communications satellite launched for the Army and Air Force, blew up a few minutes after launch. To cap this eventful week, on Friday, August 19, George Metcalf, of GE, the Advent satellite contractor, communicated GE's interest in developing a commercial launch and data-acquisition service—including a Pacific Ocean launch site.[78]

On September 1, 1960, Dr. Robert C. Seamans Jr. was sworn in as associate administrator of NASA. Seamans was born in Salem, Massachusetts, on October 30, 1918, to an old New England family. He attended Harvard in the same class as John F. Kennedy and pursued graduate studies at MIT, where he quickly became associated with Charles Stark Draper and the Instrumentation Laboratory. From MIT he went to work for RCA as founder of their Airborne Systems Laboratory. The very afternoon he joined NASA, Seamans participated in a meeting with Glennan, Johnson, and Nunn to discuss the organization of the NASA communications satellite program. Nunn was assigned the responsibility of developing public policy, international relations, and possible legislation for an *operating* satellite system. Nunn would now report directly to Glennan as special assistant for communications satellite policy.[79]

Robert G. Nunn was born in Cape Girardeau, Missouri, in 1917. He received an A.B. degree from DePauw in 1939 and a J.D. degree from the University of Chicago in 1942. After four years in the Army during World War II and eight years in private law practice, Nunn went to work for the Air Force in 1954. He became NASA assistant general counsel in 1958.[80] Nunn became the NASA communications satellite policy expert. Although he appears to have disagreed with Webb later, he would represent NASA on most of the communications satellite policy committees. Jaffe, the nominal head of the technical area of satellite communications, seems to have been peripheral to the policy process. To Glennan, and later to Webb, communications satellite policy was more important than communications satellite technology.

On September 15 AT&T's George Best and William Baker met with Glennan to provide more background on AT&T's interest. The corporation was prepared to spend $30 million for three satellite flights—more flights if they had any success. This was the first industry proposal that Glennan had received in which company funds, rather than government funds, were to be committed. A week later Glennan discussed an upcoming trip to BTL for an Echo demonstration that BTL was putting on for the FCC. His deputy and his lawyers felt this would give the appearance of NASA support for AT&T. Glennan was not amused as he wrote: "AT&T is going to be in the business and if we are going to take leadership in getting this program off the ground, it seems to me that we have to take a positive rather than a negative viewpoint in manners of this kind." At the Holmdel demonstration on September 22, a picture of Glennan and the six FCC commissioners was transmitted by facsimile to the Naval Research Laboratory and then back to Holmdel via Echo reflection. Glennan was impressed.[81]

Glennan had assigned Nunn and Jaffe the task of putting together a proposed NASA communications satellite policy and development program. He was planning on using a speech in Portland, Oregon, as an opportunity to state his view of communications satellite policy and outline a program to bring this new technology to the people of the world. Nunn and Jaffe did not prepare what Glennan wanted, so the Portland speech on October 12 was limited to policy matters. In the speech, Glennan pointed out that communications had always been a function of private industry in the United States and that NASA was proposing to provide launch services, at cost, to those companies willing to pay their own way—such as AT&T. The day after the speech, Glennan briefed President Eisenhower by telephone on what he had committed NASA to do. On October 24, becoming a little more concerned about proceeding without formal White House and cabinet approval, Glennan attempted to brief General Wilton B. Persons (White House staff) on the program and promised to provide the cabinet with a formal briefing on November 11.[82]

AT&T was back in Glennan's office on October 21. Glennan requested that a complete description of the proposed AT&T program be provided rather than the continuing parade of AT&T personnel describing the program piecemeal. This description arrived on October 24. On October 27 Glennan, Dryden, Seamans, Nunn, and Jaffe met with Jim Fisk and George Best of AT&T. NASA had analyzed the proposed program in the intervening days and was well prepared. Glennan felt that AT&T did not understand the difficulty involved in making a launch vehicle available: there were only so many rockets, and long lead times were required to establish schedules and priorities. Glennan also pointed out that other companies were interested and that the program would not be delayed for anyone.[83]

Other companies *were* interested. In addition to the Hughes visit and the many AT&T visits, ITT and Philco came to talk about communications satellites. ITT representatives met with Nunn and Lieutenant Colonel R. E. Warren on October 13 while Glennan was on his way back from Portland. ITT manufactured TWTs and had built a ground station for Courier. Although it had never built a satellite, it had the capability. ITT favored the geosynchronous twenty-four-hour satellite, for which it had budgeted $200,000 for repeater (payload) R&D in 1960. At a lower level of effort, it had budgeted $100,000 for stabilization (attitude control) R&D. ITT felt that it could design and build a satellite in three years. An operational system could be launched in 1965 for about $30 million. A communications repeater package could be made available gratis for a NASA launch. For Philco, J. Hertzberg, O. Simpson, and G. Moore, from the company's Western Development Laboratory, met on October 25 with Nunn, Jaffe, Sanders, Warren, Arnold Frutkin, and E. M. Shafer to make themselves available for future contracts. As the contractor for Courier, Philco had the advantage of being the only company that had actually built a communications satellite.[84]

Somewhat earlier, NASA had received the first report from RAND on the technology and economics of communications.[85] At the end of October, Glennan chose STL to be the systems engineering contractor for the NASA communications satellite program. It would manage the actual builder. Glennan now had the technical and economic justification for his program and a contractor to write a specification and manage the program. The stage was set.

November came and went with no sign of the cabinet paper that Glennan had been thinking about since July and had promised for the November 11, 1960, cabinet meeting. The waning Eisenhower administration was beginning to wrap up existing activities and showed very little interest in starting anything new. Generals Persons and Andrew Goodpaster chided Glennan and told him that the president wanted to mention communications satellites in his State of the Union message. On December 7 Glennan, Nunn, and Johnson met with AT&T's Frederick R. Kappel, Paul Gorman, Best, and Fisk at 195 Broadway. The NASA executives tried to suggest to the AT&T representatives that AT&T might be better off minimizing its role in satellite communications to avoid monopoly problems. Provision, at no cost to the government, of ground stations for the upcoming satellite program would be a good start. The next day at a NASA meeting on the subject, Abe Silverstein, director of space flight programs, objected strongly to the presence of private companies in the communications satellite business. Glennan was amazed and annoyed.[86] Another NASA-AT&T meeting, this time in Washington, included the details of AT&T's program. Glennan was impressed with the program, but by this time he was beginning to believe that the Washington establishment was too anti-business to allow an AT&T monopoly to occur. After presenting Eisenhower with a copy of the communications satellite briefing paper on December 19, Glennan presented the paper to the whole cabinet on December 20. Eisenhower released the details of the paper at the end of the month as his communications satellite policy—emphasizing the traditional private nature of the telecommunications industry.[87]

Nunn, the special assistant for communications satellite policy, was not a political appointee and remained at NASA into the Kennedy administration, as did Associate Administrator Seamans, General Counsel Johnson, Communications Program Manager Jaffe, and many others. While most of the administration were packing their bags and Glennan was enjoying his Christmas holiday, Nunn and Johnson met with Attorney General William Rogers on December 23 to discuss NASA's future relationship with AT&T. Nunn pointed out that NASA's problem consisted of the facts that, on the one hand, AT&T was "realistically the only company capable of doing the job" and that, on the other hand hand, the monopoly power of AT&T and its attempt to "preempt" the role of communications satellite builder and operator "would in effect select AT&T as the 'chosen instrument' of the United States." Rogers made two points: (1) the government should not act to put AT&T into a preemptive position, nor must it appear to do so; and (2) "the Executive Branch probably should obtain at least the acquiescence of Congress." Nunn and Johnson showed Rogers the Glennan position paper that would be released by the White House. Rogers had no general problem with the paper but objected to two sentences that specified "private enterprise" as the system operator and rejected government operation of a communications satellite system. Even so, the sentences remained in the statement.[88]

Satellite communications systems were generally seen to be something that private industry should operate, but by 1960 there was a growing question as to which agency was responsible for communications satellite policy. The answer depended on whether communications satellites were seen as part of telecommunications policy (FCC) or space policy (NASA). By 1960 AT&T was already committed to building and operating an MEO (medium Earth orbit) active communications satellite. The satellite was designed but not built. Hughes Aircraft Company had actually built a prototype model of its geosynchronous satellite, but not being a communications company, it was reluctant to launch the satellite without partners. Because of its monopoly position as a communications seller, AT&T was also the only potential buyer of communications satellites.

The drama of the space race may have led to NASA's being handed responsibility for communications satellite policy. The FCC would seem to be the logical choice to develop satellite communications policy if industry was expected to develop the technology. Clearly many civil servants, especially at NASA, felt that this was something the government should do. Part of the rationale seems to have been the limited number of available launch vehicles. Most of these would be used for reconnaissance satellite launches. NASA would want to use the rest. Acting as "gatekeeper" was a not-unreasonable task for NASA.

No decision reached in 1960 seems to have excluded AT&T, but several trends were already visible. Many NASA civil servants (e.g., Abe Silverstein, director of space flight programs) felt that NASA should develop—and possibly operate—commercial communications satellites. Political appointees (e.g., Attorney General Rogers) were uncomfortable with the idea of "helping" AT&T. Others (e.g., NASA Administrator Glennan) saw AT&T as the obvious choice to operate a satellite communications system. Foreign policy issues may also have risen at this point. AT&T had been negotiating with foreign PTT (Post, Telegraph, and Telephone) administrations for a half-century, but suddenly the State Department wanted to control these negotiations. Congress had been interested in the issue of satellite communications for several years but did not seem to reflect a single opinion. Many members of Congress simply assumed that AT&T would operate the commercial satellite communications system. Others thought that the initial system should be government controlled.

Hughes had a tremendous advantage in having developed a lightweight geosynchronous satellite. Unfortunately, GEO was the exclusive domain of military satellite communications, and the military already had an ongoing program: Advent. Hughes could not sell to AT&T, so it would build its own satellite. Hughes was trying to form a partnership to circumvent the AT&T monopoly but was having little success. Its main hope was to persuade NASA to support the further development of the Hughes satellite. This support should probably continue through to operational status, since the AT&T monopoly would otherwise remain a problem.

No clear path was obvious, but only AT&T seemed to have any chance of operating a commercial communications satellite system. It would soon submit a proposal to NASA expressing a willingness to cover all costs. Hughes would also submit a proposal to NASA, but it would do so reluctantly because NASA could support only MEO systems. GEO systems were reserved for DoD.

4. Government Intervenes

John Logsdon's study of the U.S. decision to land a man on the Moon has shown how emerging space technology was utilized as an instrument of national policy to achieve national objectives. Inasmuch as Kennedy included an accelerated [communications] satellite program in his man-on-the-moon message to Congress, the same can be said for U.S. [communications] satellite policy. —Jack Oslund (Comsat), "Open Shores to Open Skies," 1977

Why should NASA develop an operational system? —Robert G. Nunn (NASA assistant general counsel),1961

John F. Kennedy's preelection attitudes toward the space program are not clear,[1] but his transition team had developed some strong feelings about the space program in general, and about satellite communications in particular. Jerome Wiesner, the new presidential science adviser, had been given the task of examining the U.S. space program and making recommendations to the president. Wiesner had been director of the MIT Radiation Laboratory and had served on Eisenhower's PSAC (President's Science Advisory Committee). Wiesner was critical of the space program in general; in particular, he felt that the development of a communications satellite system was beyond the investment capabilities of industry.[2] Wiesner had proposed a joint U.S.-Soviet communications satellite program in 1959.[3] Buried in this recommendation was his feeling that this new technology should not become an AT&T monopoly. At least some members of the Kennedy administration were prepared to overturn the Eisenhower attitude that private industry was the logical candidate to develop commercial satellite communications. They would find allies within NASA, within the State Department, and within Congress.

Intervention by NASA

On January 4, 1961, the Request for Proposal (RFP) for the NASA experimental communications satellite was released. On January 11 Glennan met again with Jim Fisk, of AT&T. Fisk promised to build a ground terminal in the United States and make it available for communications satellite testing. In addition, Fisk would negotiate with AT&T's European partners for a ground terminal on the other side of the Atlantic.[4] Further discussions with NASA personnel on the January 11 and 12 made it clear that the civil servants were against AT&T's involvement. Glennan wrote in his diary, "[Leonard] Jaffe and [Abe] Silverstein seem determined that anything short of having someone other than AT&T win the competition will be tantamount to following a 'chosen instrument' policy."[5] Glennan had done his best, but he was leaving behind civil servants who seemed strongly prejudiced against private industry.

The combination of the critical Wiesner report and the lack of contact between the new administration and NASA management made everyone at NASA very nervous. Glennan had generated a turnover file and had made known his interest in effecting a seamless transition, but nobody called. NASA Associate Administrator Robert C. Seamans Jr. and General Counsel John Johnson speculated that some of NASA's responsibilities might be transferred back to the military.[6]

On January 30, 1961, President John F. Kennedy nominated James E. Webb to be the new NASA administrator. On February 9 Webb was confirmed by the Senate and on February 14 took the oath of office. Webb may have been Kennedy's second choice after General James Gavin, but Webb did fit the requirements. What was needed was a political manager, not a scientist and not an engineer. Glennan's background was technical, but he had also been a manager almost all of his career and had a deep knowledge of the interplay of politics and technology. Webb had a less technical background, his only claims being his experiences as a U.S. Marine pilot in the 1930s and as an employee of Sperry Gyroscope, but he had better political credentials than Glennan and equally impressive management credentials. Born in North Carolina and trained as a lawyer, Webb had been director of the Bureau of the Budget (BoB)[7] and undersecretary of state in the Truman administration. He had worked for the Kerr-McGee Oil Company and had been recommended to Lyndon Johnson by Senator Robert Kerr (D, Tex.).[8]

Kennedy's State of the Union address had included an invitation for all nations to join in a new communications satellite program. Both AT&T and ITT had offered to provide international ground stations for active communications satellite experiments—at no cost to the government. AT&T's offer was accepted for ground stations in the United States, but NASA felt that the State Department, rather than AT&T, should make arrangements for foreign stations. Although AT&T had already cleared the way for international experiments, NASA made formal agreements with the United Kingdom (on February 14, 1961) and France (on February 16, 1961) to participate in the testing of Relay and Rebound. On February 27 NASA and the Federal Communications Commission (FCC) signed a memorandum of understanding that defined their respective positions and responsibilities. The FCC would still be responsible for spectrum allocation, but NASA would be responsible for technology and policy.[9] Webb met briefly with NASA Assistant General Counsel Robert Nunn on February 27 and asked for a briefing on communications satellite developments. He also suggested that Nunn prepare a policy paper on these developments. Nunn responded the next day, emphasizing that NASA had a policy formation role. He suggested that Webb read a United Research report favoring government ownership of communications satellites and the Glennan cabinet paper favoring private ownership.

During March NASA received seven proposals—from AT&T, Bendix, Collins Radio, Hughes, ITT, Philco, and RCA—to build the low-altitude Relay satellite. During this period, NASA was preparing its fiscal year 1963 (FY63) budget estimates. The budget guidelines contained several assumptions: (1) no funding of operational systems (demonstrations only), (2) no funding of ground support, and (3) no development of passive satellites. Approved programs included two Thor-launched Relay satellites and two Atlas-Agena-launched Relay satellites.[10]

The NASA FY62 budget proposed by the outgoing Eisenhower administration included $34.6 million for communications satellite development. There had been some controversy as to the amount that industry should reimburse NASA for its communications satellite development. A rather arbitrary $10 million industry contribution had been included in the communications satellite program plan. This was specified in the RFP released in January 1961. In December 1960 Glennan had asked Maurice Stans, director of BoB, to put the $10 million industry contribution back in the NASA budget so that the government would not be dependent on industry. Stans refused. In a January 14, 1961, press conference on the NASA budget, Deputy Administrator Hugh Dryden was asked why industry was not paying more than $10 million, given its obvious interest. Dryden made a noncommittal response. In February Webb, Seamans, and Senator Kerr discussed the $10 million, with Kerr and Webb agreeing that it should be in the supplemental budget. This $10 million finally showed up in the March amendment.[11]

Webb was anxious to make NASA move quickly in order to beat the Soviets. In a March presentation to President Kennedy, Webb portrayed the Eisenhower administration as having "emasculated" the civilian space program's ten-year plan (1959–69) by reducing the FY62 NASA budget by $240 million. This reduction "guaranteed" Soviet space superiority for the next five to ten years. Webb allowed that the low space budgets of the Eisenhower administration had permitted "extensive" scientific investigations, development of satellite applications (weather and communications), and the Mercury program but had offered no opportunity to get ahead of the Soviets. Webb's most heartfelt plea was for an expanded booster program. At the center of this program was the Saturn launch vehicle, consisting of eight clustered Atlas engines producing 1.5 million pounds of thrust, followed by the Nova launch vehicle with 6 to 9 million pounds of thrust. These vehicles, as well as NASA's Centaur stage for the Atlas, would also be available for military missions, such as the Advent communications satellite. This could be done by gradually increasing the budget to $2.0 billion in 1965. This increased budget would produce large boosters capable of manned trips to the vicinity of the Moon. In addition, a new era of international cooperation in weather and communications satellite systems could possibly ensue, but only if the United States was seen as a leader in space science and technology.[12]

Webb, possibly in response to administration attitudes or to comments by NASA civil servants, was taking a dim view of AT&T's communications satellite program. In April Fred Kappel, president of AT&T, and Webb exchanged a strange series of letters. In an April 5 letter Kappel complained that Webb had stated that NASA had "yet to receive a firm proposal from any company" to form a communications satellite development partnership with NASA. Kappel recapped the basic facts of AT&T's communications with NASA over the past year and pointed out that in the Relay proposal, AT&T had volunteered to share costs and even to contract privately for rockets and launch facilities.[13] Webb replied, on April 8, in a rather unfriendly tone: "I am told that your letter of December 14th was delivered by a number of your associates [Jim Fisk and George Best], that an extended conference ensued, and that it was made clear that NASA would not permit your company, or any other, to pre-empt the program of the United States in this area."[14] Glennan's response to the December 14 meeting had actually been that AT&T had proposed "a rather good program."[15] The battle lines were being drawn, and AT&T was on the losing side.

Meanwhile, the cold war was heating up. On April 12, 1961, Soviet Major Yuri A. Gagarin made man's first orbital flight. A few days later the Bay of Pigs debacle ran its unfortunate course. There were many reactions to these events. One was an urge to do something spectacular in space before the Soviets. Some suggested that space was the realm of the military—not NASA and civilians.[16] On April 20 President Kennedy asked Vice-President Lyndon Johnson, in his role as chairman of the National Aeronautics and Space Council (NASC), to recommend a program in which the United States could beat the Soviets. On April 28 Johnson replied that the United States could probably beat the Soviets in the race to the Moon.

On this same day Webb was complaining to the NASA general counsel about AT&T's contacts with other international carriers regarding satellite communications. Webb also discussed his feeling that an early determination of public policy often led to unhappy results.[17] A few days later Webb wrote to the chairman of the House Committee on Science and Astronautics and suggested that the upcoming hearings on satellite communications were premature given the imminent FCC determination of satellite communications policy as a result of responses to FCC Docket 14024 and the imminent award of the NASA Relay contract.[18] At the same time Wiesner, Kennedy's science adviser, had a panel investigating the "technical" issues involved in satellite communications. Nunn was keeping track of this panel's investigation. The panel seemed to be puzzled over the NASA-DoD split but felt that the civilian-military and government-industry roles were political decisions, although it recognized that operating a global communications satellite system would be easier for an existing international communications industry than it would be for a new government entity.[19] On May 8 AT&T President Kappel wrote another letter to Webb, reminding him that AT&T was ready to launch a satellite at its own expense and would deliver this satellite to NASA for launch eight months after being told what launch vehicle would be used.[20]

Department of Defense

Advent was still in trouble. In March the program was reviewed, and a report was issued in April.[21] After almost three years of effort, the Advent program was in disarray. Management problems were so bad that technical problems were not completely visible and were not attacked in a coherent manner. The Air Force was responsible for the launch vehicle and the satellite (called FSV: final stage vehicle) while the Army was responsible for the communications payload and the communications ground system. The Army had been given overall responsibility for defense satellite communications in September 1959, but overall management authority for Advent had not been granted to the Army until September 1960.[22] The Army Advent Management Agency (AAMA), under the command of Brigadier General W. M. Thames, was created to perform this program management function at U.S. Signal Corps facilities in Fort Monmouth, New Jersey.

Those rather convoluted chains of command meant that Space Technology Laboratories (STL) systems engineers hired by AAMA had to request AAMA to forward to the Air Force their request for information on the launch vehicle, an Atlas being built in San Diego by Convair, and for information on the satellite being built in Valley Forge by General Electric (GE). The Air Force had hired Aerospace Corporation to perform systems engineering tasks. Aerospace Corporation engineers had to request the Air Force Ballistic Missiles Division to forward to the Army their request for information on the satellite communications payload (Bendix) and ground stations (Sylvania). The Army was responsible for communications ground stations, but the Air Force was responsible for the tracking, telemetry, and command (TT&C) ground stations being built by Philco. When pointing and positioning requirements of the spacecraft were being discussed by the Air Force, it was not always possible for the Army to ensure that communications requirements were being met. When spacecraft power was allocated, it was not always possible for the Army communications engineers to ensure that their needs were met. Both spacecraft "bus" and payload increased in weight, but Atlas-Centaur capability actually decreased. Trade-offs and compromises were difficult to optimize because neither the subsystem experts nor the systems engineers (STL for the Army, Aerospace for the Air Force) could talk directly to each other. The DoD was deliberating among various options: continue, terminate, or modify the design, the management, or both. The basic argument for termination was that the probable $500 million cost of the system was too much for its limited capabilities. The counterargument was that this was the only DoD communications satellite program left at a time when placing a satellite communications system in operation was becoming a more urgent priority.

Congressional Hearings: May 8, 1961

On January 19, 1961, the FCC had granted AT&T's request for authorization to launch an experimental communications satellite system. The only previous significant communications satellite action by the FCC had been the reopening of Docket 11866 (known as "above 890") in May 1960 (Docket 13522, modified in December), soliciting opinions on the frequencies required for space communications. In response to what appeared to be a rush to reach a decision in the satellite communications arena, the FCC opened Docket 14024 in March 1961, soliciting views on the "administrative and regulatory problems" associated with a commercial satellite communications system. A little more than a month later, on May 8–10, the House Committee on Science and Astronautics held hearings on communications satellites. Many of the witnesses included their responses to the FCC's Dockets 11866/13522 and 14024 as addenda to their testimony. The committee chairman, Overton Brooks (D, La.), stated in his opening remarks: "The proper relationship between Government and Industry must be defined . . . the most desirable business arrangements should be determined at the earliest possible time."[23]

In its May 1, 1961, response to FCC Docket 14024, RCA emphasized the benefits of a joint venture and the disadvantages of a monopoly. RCA was still unclear as to its own interest in participating in such a joint venture, but it did intend to own and operate Earth stations to provide satellite communications services. Dr. Elmer W. Engstrom, senior executive vice-president, made RCA's presentation to the committee. He pointed out that RCA, GTE, and Lockheed had teamed to study satellite communications. Engstrom described a geosynchronous system of two to three satellites in equatorial orbit 42,000 kilometers from the center of Earth. Each satellite would weigh approximately 350 kilograms. The most important use of such a communications satellite system would be public-service television—especially telecasts of United Nations meetings. The cost would be in the hundreds of millions of dollars, perhaps as much as a half-billion dollars. RCA seems to have been aware of the technical advantages of GEO (geosynchronous Earth orbit) but had not recognized the commercial value of communications satellites.

RCA's response to FCC Docket 13522 was also included in this presentation to Congress. Its March 1, 1961, response suggested that frequencies in the 800-to-10,000-MHz range be set aside for satellite communications. In the long run, additional frequencies above 10,000 MHz might also be needed. RCA stated to the FCC that it could not justify a satellite for its own purposes but that the aggregate needs of all the international carriers might justify the expense of a satellite. Included in the Docket 13522 response was the description of a geosynchronous communications satellite designed by Sidney Metzger of RCA Astro-Electronics. The satellite used single-sideband (SSB) modulation for the uplink and frequency modulation (FM) for the downlink. It was three-axis stabilized to allow continuous pointing of an antenna at Earth while the solar arrays tracked the Sun.

Engstrom suggested, in concluding his testimony, that three principles should guide the direction of U.S. communications satellite policy:

1. Regardless of the ownership of space satellites, all international communications common carriers should have equitable and direct access to, and nondiscriminatory use of, the satellites.
2. The satellites should be available to such carriers on reasonable terms to use for any services which the Federal Communications Commission au-

thorizes them to provide now or in the future, without any restrictions imposed against such use, through contract or otherwise, by the owner or other organization controlling the satellites.

3. Each such U.S. carrier and overseas communications agency should have the right to establish, own, and operate its ground stations for transmitting and receiving signals via the satellites.[24]

It seems probable that RCA was unsure about satellite communications. At some levels satellite communications was viewed as a government demonstration of the peaceful uses of space—in which case RCA might want to build the satellite. At the very least, RCA wanted rights similar to those given to it in the telephone cables that AT&T had pioneered in laying across the oceans.

GTE was represented by Theodore F. Brophy, vice-president and general counsel. After an introduction, which included a mention of GTE-Sylvania's contract to build the Advent Earth stations, Brophy expressed GTE's interest in participating in a jointly owned and operated geosynchronous satellite system. Like RCA, GTE included its responses to Dockets 13522 and 14024 in its presentation to Congress. GTE's March 21, 1961, response to Docket 13522 was very similar to RCA's response. The major difference was a greater eagerness on GTE's part to be a part-owner of the satellite system. GTE noted that only one system was likely and that if the one system was geosynchronous, it would make participation by all carriers more likely. In its response to Docket 14024, GTE made clear that it expected to participate in a satellite company "owned by all existing and future domestic and international U.S. communications common carriers."[25]

In response to questions from Joseph E. Karth (D, Minn.), who was temporarily presiding over the committee meeting, Brophy differentiated the GTE position from the RCA position. First, GTE was convinced that the satellite ownership problem had to be addressed quickly to assure potential participants that their rights would be protected; specifically, GTE was concerned that domestic carriers be included as well as international carriers. Second, GTE argued that the communications capability of the satellite be allocated on a demand-assignment basis. Instead of permanent ownership of voice channels, frequencies, or transponders, each Earth station would be assigned a channel as need arose—for each individual phone call. Brophy concluded by stating that he thought the half-billion dollars needed for funding a communications satellite system could easily be raised in the stock market.

Like GTE, ITT fully intended to participate in ownership and operation of satellite communication facilities. Henri Busignies, vice-president and general technical director, made the ITT presentation. After an outline of ITT's business areas, a recapitulation of the 1959 ITT testimony, and a tutorial on the problems, Busignies described ITT's current conclusions. Like RCA and GTE, ITT favored a geosynchronous satellite. Busignies made the case for FM (lowest power) rather than SSB (narrowest bandwidth) in the satellite downlink. The uplink was not described. It should have been obvious at this point that use of different modulation techniques for uplink (SSB) and downlink (FM) would require more complexity on the satellite, since the uplinked signal would need to be demodulated and then remodulated, but none of the participants seem to have considered this.

In its response to Docket 14024, ITT's view was that only international communications common carriers should participate in the satellite communications enterprise. It pointed out that overseas companies and administrations would also participate. As RCA and GTE had done, ITT gently raised the specter of monopoly. ITT also brought up the issue that was later described as "one phone call, one vote"—suggesting that shares in the enterprise be in proportion to expected use of the system.

Hillard W. Paige, general manager of the Missiles and Space Vehicle Department, made the GE presentation. The GE presentation and copies of GE's FCC filings take up more pages in the Hearings document than for any other presenter—50 percent more than AT&T. After a short overview of GE's contributions to space technology, including its manufacturing of the DoD Advent communications satellite, Paige presented GE's view: a space communications system should be established quickly, by private industry using private funds. The system should be a joint venture of the communications and aerospace industries. GE had already established Communications Satellite (Comsat) Corporation, which could be the vehicle for the joint venture. GE proposed an equatorial, 6,000-mile orbit for the system. The design was generally similar to that of the Advent system, which GE was already building.

GE's response to Docket 14024 was to propose that the aerospace and communications industry jointly raise $250 million to put the GE medium-altitude equatorial system of ten satellites in place, using the Atlas-Agena launch vehicle. Another $250 million would be needed to build the ground system. The communications industry might have sufficient expertise to build the ground system (Earth stations), but the aerospace industry had the expertise to build the satellites. Like ITT, GE pointed out the necessity of foreign participation and the apparent capability to support only one commercial communications satellite system. And like the other presenters except AT&T, GE pointed out the necessity of avoiding monopoly. Government support would be required—but not government funding. GE was ready to put $25–$50 million of its own money into Comsat.

Table 4.1
GE Satellite Mass by Subsystem

Subsystem	Mass (pounds)
Attitude Control	160
Orbit Control	58
Tracking and Telemetry	15
Electrical Power	439
Communications Electronics	240
Antennas	10
Structure and Thermal	80
Total	1,002

In its Docket 13522 response, GE recommended the same range of frequencies suggested by most of the other presenters: 1–10 GHz. In its testimony, and in the description of the GE system, GE stressed the ground-microwave frequencies in the 6- and 4-GHz band—the frequencies originally rejected by AT&T in the "above 890" controversy. The description of the GE satellite is very similar to the Advent parameters. The subsystem weights are given in Table 4.1.

The 1,000-pound, 700-watt GE satellite would provide 300 voice channels and a television channel. Transatlantic telephone cables had first been constructed in 1956 at a cost of over $1 million per voice channel, but costs were now closer to $500,000 per voice channel. The five satellites of GE's proposed interim system, including ground stations, would be worth about $750 million at this cost. The final ten-satellite system would be worth about $1.5 billion. AT&T was, however, already envisioning submarine cable voice channels as cheap as $100,000 each. The very large satellite proposed by GE was cheaper than cable, but not necessarily for long. Satellites could also transmit television signals, but that market was not well defined.

James E. Dingman, vice-president and chief engineer of AT&T, made the next presentation. Most of his experience was with the Long Lines Department, but he had also been general manager of BTL. Dingman started his presentation by stating that it was possible to "build and launch an experimental communications satellite within 9 or 10 months from go-ahead, and to have a commercial system in operation with[in] 3 to 4 years."[26] Dingman argued that demand for overseas telephone calls was increasing at a rate (20 percent per year) that exceeded the ability of cable systems to supply. "One or more satellite systems" would be needed to supplement and complement the cable systems. In addition, he noted: "It would be a severe blow to [U.S.] prestige should another country take the lead." AT&T again proposed its system of fifty satellites in medium-altitude (7,000 miles) inclined orbits. AT&T was ready to put such a system into operation, at its own expense, if only NASA would sell it launch services.

In its May 1, 1961, response to FCC Docket 14024, AT&T proposed that its satellite system be a joint undertaking of all the international communications common carriers, foreign and domestic. The ground stations could be jointly owned or separately owned, or channels could be leased with ground-station facilities included. Also included in the Hearings document of the AT&T May 10 testimony is the company's May 15 reply to the other respondents to Docket 14024. AT&T specifically challenged the benefits of including companies that were not in the international communications business (e.g., Lockheed and GE) and suggested that a new satellite communications corporation would not be of any benefit and might do considerable harm.

In response to questions following his prepared statement, Dingman estimated that the entire satellite system would cost several hundred million dollars. Dingman estimated that AT&T and its overseas partners had invested about $275 million in the existing cable system. He felt that the initial system would consist of a few 150-pound experimental satellites launched on Thor vehicles, with later satellites growing significantly in weight and being launched on Atlas vehicles. The experimental system could be launched in 1962 and the operational system in 1964. In response to questions about frequencies to be used, Dingman affirmed that discussions with foreign partners had made clear that the common-carrier band in the United States—3,700–4,200 MHz and 5,925–6,425 MHz—was probably the easiest band on which to establish satellite communications, contrary to AT&T's original "above 890" comments.

During the question-and-answer session, Representative James G. Fulton (R, Penn.) asked about AT&T's December 1960 proposal to NASA. AT&T responded that it had been waiting since December 14, 1960, for NASA to agree to sell it launch services. Brooks, the committee chairman, was also somewhat annoyed that NASA had not responded: the committee had been discussing this issue internally since October. NASA claimed to be unready to testify but said that the agency would come before the committee in a "couple of weeks" (it would actually be two months).

L. Eugene Root, group vice-president for missiles and electronics, made the Lockheed presentation. Root was accompanied by Beardesley Graham, special assistant on communications satellites. Lockheed's interest in space dated back to at least 1954 and was closely tied to the Air Force. Shortly after the launch of *Sputnik 1*, Lockheed had been formally awarded the contract for the Air Force WS-117L satellite (Samos/Midas). Because the Air Force satellite was moving slowly, an "interim" CIA design was implemented: the Discoverer (Corona) reconnaissance satellite. Lockheed also built this satellite and the Agena upper-stage common to both programs. The Agena was built to provide in-orbit propulsive capability for the reconnaissance satellite program. Thor-Agena variants launched all but 3 of 143 Discoverer satellites over the next dozen years (so much for "interim"). Agena upper-stages were dedicated almost exclusively to reconnaissance satellites until 1971 and the advent of the Big-Bird satellites, which were launched on the Titan. The Agena stage was used to launch only three communications satellites: *ATS-1, ATS-2,* and *ATS-3* in 1966–67.

Root testified that Lockheed's Air Force work had convinced the company that the problems of commercial communications satellites were not technical but rather legal, regulatory, economic, and international. In late 1959 Lockheed began work on these issues and in early 1960 hired Booz, Allen & Hamilton to study the issues. Lockheed even responded to FCC Docket 11866 ("above 890"). Late in 1960 RCA and GTE joined the Lockheed study team.

Lockheed's response to FCC Docket 14024 was to propose the creation of a new organization, Telesat, which would operate the global communications system. Telesat would be owned by the communications carriers, other companies, and the general public. Government subsidies would be advisable in its early years to reap the prestige benefits of inaugurating a global communications satellite system. There would be no foreign ownership of Telesat, but foreign organizations would own their own ground stations and might receive an undivided ownership interest in the satellites, but not of the Telesat company itself. The satellites would be in GEO, two over the Atlantic and two over the Pacific. R&D costs would be about \$100–\$150 million, the initial operating system would cost about \$50–\$65 million, and an additional \$50–\$100 million would be required in the early operational period. The \$200–\$315 million system would not be self-supporting until the mid-to-late 1970s.

In the discussion following Root's formal presentation, several points were made, including the availability of launch vehicles and the cost of different systems. The available launch vehicles—Thor, Atlas, and Titan—were all based on military vehicles. Root felt these were all committed to government missions. Graham told the committee that the cost of the four-satellite geosynchronous system and the cost of the fifty-satellite medium-altitude system were both in the \$200–\$300 million range.

One of the common themes in the testimony of all the companies, with the exception of AT&T, was the problem of monopoly—AT&T was not usually mentioned by name, but everyone knew the target. Another common theme was GEO, which was preferred by ITT, RCA, GTE, and Lockheed. AT&T, ITT, and GTE, all communications companies, thought that communications satellites would pay for themselves. GE agreed, but RCA and Lockheed saw the need for some form of subsidy. Interestingly, NASA felt that these hearings should not even be held. The implication seems to have been that NASA's choice of a Relay contractor would determine the future course of the technology. This was not to be the case.

NASA's Choice for Relay

On Friday, May 12, shortly after the closing of the House hearings at NASA's request, Seamans presented Webb and Dryden with a summary of the accelerated NASA program proposed in response to Kennedy's desire to beat the Soviets.[27] Seamans had worked with John Rubel, the deputy director of defense research and engineering, on the combined NASA-DoD response over the previous weekend.[28] Three specific areas were addressed: (1) manned lunar landing, (2) communications satellites, and (3) meteorology. In a May 16 memorandum to Seamans, Nunn referred to Seamans's memorandum of May 10 and raised some basic issues revolving around the following question: "Why should NASA develop an operational [communications satellite] system?" This question was critical to Nunn, since the communications industry continued "to affirm its own clear intent and obvious ability to achieve the same objective."[29]

On May 18, 1961, RCA was awarded a contract to build Relay. There was some irony in this choice: RCA had publicly stated its preference for the twenty-four-hour satellite. Before announcing the RCA win, Webb called Kappel, president of AT&T, to tell him that NASA would sell him launch services. Although NASA did not announce the standings, the ranking was apparently (1) RCA, (2) Hughes, (3) Philco, and (4) AT&T. ITT, Bendix, and Collins Radio were apparently far behind these four. Some NASA participants in the evaluation process expressed surprise; they had expected the AT&T proposal to be better, if not the best. The deciding factor was apparently the RCA 10-watt traveling-wave tube. On May 24 the FCC announced that it favored a communications satellite system jointly owned by the international communications carriers. On May 25 President Kennedy made the speech in which he challenged the nation to commit to a manned Moon landing within the decade. Included were the other two goals that NASA had suggested—communications and meteorology from space—and a fourth goal, the building of a nuclear rocket. In statements and briefings that day, NASA officials indicated that the AT&T satellite would also be launched, but they did not clarify what they would do with the additional $50 million that Kennedy had placed in the NASA budget for satellite communications. At the administrator's staff meeting that day, Dryden explained that the additional money was not for the purpose of a separate technical approach but rather for speeding up development. Fortunately for Hughes Aircraft Company and its small geosynchronous satellite, this was not to be the case.[30]

At the June 1 administrator's staff meeting, NASA's top international official, Arnold Frutkin, reported on the trip that he and Leonard Jaffe had taken to Brazil. It was their impression that ITT would be providing an Earth station for Relay experiments.[31] On this same date, Webb assured McNamara that he would be kept fully informed of all communications satellite activities.[32] At the next staff meeting, Nunn reported on the June 5 FCC meeting to form the ad hoc carriers committee. This committee would be tasked with forming a communications satellite joint venture consisting of all the international communications carriers. GE had filed a petition to reconsider the exclusion of noncarriers. Seamans and Abe Silverstein, director of the Office of Space Flight, discussed the technical meetings that were about to take place with AT&T. Certain policy problems were bound to arise. They agreed that there needed to be a report at the end of AT&T's experimentation and that a procedure for handling relationships between AT&T and foreign PTT (post, telegraph, and telephone organizations) Earth-station operators had to be developed. Dryden stated that AT&T would have to agree to allow NASA to control international arrangements and would also have to agree to use the Relay ground stations. In addition, the question of patent rights and reimbursement had to be settled.[33]

On June 12 several delegates from GE met with Webb. They had used the good offices of Senator Hubert Humphrey (D, Minn.) to obtain an appointment. They argued their position that noncarriers should be part of the joint venture. They stressed that they had been in touch with Dr. Edward C. Welsh, of the NASC, who agreed with them that the FCC was undertaking more responsibility than it should. Webb countered with a defense of the FCC position and an indication of the two programs that NASA was pursuing—the NASA-RCA Relay satellite and the launching of the AT&T satellite—as well as "new and novel" designs such as that proposed by Hughes. The GE delegates indicated that they would either withdraw their opposition to the FCC plan or submit a new position.[34] At the June 15 staff meeting, Nunn reported that the GE brief would be submitted on June 19, the FCC rebuttal on June 24, and the FCC decision on June 25. Nunn also reported that GTE was claiming to be an international communications carrier because it had communications interests in South America and the Philippines.[35]

At the June 22 staff meeting, Nunn reported that on June 15 President Kennedy had directed the NASC to study communications satellite policy. Nunn also reported that the FCC had delayed its decision on joint-venture membership until the first week of July. In discussing the arrangements with AT&T, NASA General Counsel Johnson noted that the FCC commissioner and the assistant attorney general were pleased with the NASA-AT&T patent policy. This policy gave the government not only royalty-free use of AT&T patents

Tom Hudspeth (left) and Harold A. Rosen with Hughes Syncom prototype on the Eiffel Tower in 1961. They were told, "That's as high as it will ever get." (Courtesy Hughes/BSS)

but also licensing rights. Johnson suggested that this policy be incorporated into the RCA Relay contract. Silverstein reported that the Davenport study commissioned by Wiesner, Kennedy's science adviser, had reported that both Advent and Relay were technically sound and necessary. Similarly, the DoD Clark Committee had concluded that both Advent and the Hughes satellite were technically sound and deserving of development.[36] A week later, Seamans was discussing the possibility of funding the Hughes "small synchronous communications satellite program."[37]

A meeting was held in Paris on July 3–4 to discuss Relay project operations. Present were representatives of American (AT&T, ITT), French (CNET, CGE), British (BPO), German (DBP), and Brazilian communications entities. In addition NASA, RCA, BTL, STL, and Siemens representatives were present. The first day started with a brief overview of the Telstar and Relay programs. This raised an issue with almost all of the participants: Relay apparently

Tom Hudspeth, Harold A. Rosen, and Donald D. Williams with Syncom and the Hughes lightweight traveling-wave tube in 1963. (Courtesy Hughes/BSS)

was incapable of handling two-way telephone operation. Most of the participants indicated that they were not interested in proceeding if Relay could not be modified for this purpose. Ground-station construction designs and schedules were also discussed. The second day continued the discussion of two-way telephony and raised the issue of documentation. All the communications entities agreed that too much paperwork was being required by NASA. The British representatives said they would have to reconsider their participation.[38]

In a few weeks in May and June 1961, NASA had taken over civilian development of communications satellite technology, and the White House NASC was about to take over communications satellite policy. It is not surprising that the communications entities were becoming a little restless—they still saw this as an extension of the kinds of relationships they had built during the laying of the transatlantic telephone cables. NASA and the NASC were taking a commercial communications program and making it part of the space race with the Soviet Union.

The National Aeronautics and Space Council (NASC)

The National Advisory Commission on Aeronautics (NACA), the predecessor to NASA, had been overseen by a board (the commission) that consisted of a broad range of people who brought their expertise to NACA. The board did not manage NACA but rather set the broad details of policy and program. The NASC, created by the 1958 Space Act, was meant to be an advisory board rather than an active board that set policy. The NASC met at best once a month during the Eisenhower administration. There was serious thought about abolishing the board.

Vice-President Johnson had been very active in space affairs when he had served as Senate majority leader, and he wanted to maintain that activity. In December 1960 Kennedy announced that Johnson would be chairman of the NASC (the law named the president as chairman and thus had to be changed). In March 1961 Johnson chose Dr. Edward Cristy Welsh to be executive secretary of NASC. Welsh had been legislative assistant to Senator Stuart Symington (D, Mo.), a member of the Senate Aeronautics and Space Sciences Committee. Welsh was born in New Jersey in 1909. He graduated from Lafayette College in 1930 as a journalism major. He received an M.A. in economics from Tufts in 1932 and a Ph.D. in economics from Ohio State University in 1940. During his time at Ohio State, Welsh had also worked in Washington as a New Deal economist. During the war years he worked in the Office of Price Administration (OPA), eventually becoming deputy administrator. In 1947 he went to Japan as head of the Antitrust and Controls Division. In 1950 he began a long association with Senator Symington; later he helped draft the Space Act of 1958. Welsh's experience made him somewhat partisan in politics and a strong advocate of "trust-busting."

Immediately after Welsh's appointment, he met with Kennedy, Johnson, David Bell (BoB), and Wiesner (science adviser) to discuss the space program. Much of Welsh's early correspondence is about office space and staff, but he was clearly very close to Vice-President Johnson. Early in his term, Welsh suggested that communications satellite policy was a natural for the NASC and dedicated more staff time to this issue than to any other in the next year. One of his first actions was to recommend rescinding the Eisenhower administration's requirement that industry contribute at least $10 million to the NASA Relay program.[39] During much of April 1961, Welsh was planning space policy for the administration. Welsh had scheduled meetings with the House and Senate Space Committees on May 1 and 2, but late in April he canceled these meetings. The Soviets' launch of Gagarin into orbit on April 12 and the failed Bay of Pigs invasion on April 16 had lowered the prestige of the United States and the Kennedy administration. Even after these events, as late as May 17, Kennedy invited the Soviet leader Nikita Khrushchev to join him in sponsoring a joint U.S.-Soviet exploration of Space. Khrushchev wasn't interested.[40]

President Kennedy had been planning a major speech to Congress since April. The president needed to show that the United States was not a paper tiger. Space seemed to offer that opportunity. Vice-President Johnson had suggested that the United States could beat the Soviets to the Moon. Alan Shepard's suborbital flight on May 5, 1961, suggested that perhaps the United States was actually not that far behind the Soviets. In addition to the spectacular Moon program and the large boosters, Kennedy (or his speechwriters) wanted to offer something practical. Before Kennedy's May 25 speech, Welsh immersed himself in FCC, NASA, DoD, PSAC, and industry dealings with communications satellites. Not surprisingly, many participants looked with disdain upon the efforts of other participants, but all agreed that it was a technology rapidly approaching the point of practicality. Kennedy's May 25 speech benefited from Welsh's efforts. In addition to promising to land a man on the Moon "within the decade," Kennedy committed the United States to building a global communications satellite system. Kennedy also added $50 million to NASA's budget for satellite communications. The money would be used wisely.

The reactions to Kennedy's satellite communications proposal were immediate. By June 5, 1961, Welsh had heard from the State Department, NASA, BoB, PSAC, and probably many others, all with comments and complaints about the communications satellite program. In a memorandum for Vice-President Johnson, Welsh outlined the steps he was taking and provided a list of questions that should be answered in order to establish satellite communications policy. Welsh's comments and questions make it clear that private ownership of communications satellites was problematic. The State Department wanted to emphasize communications with Third World countries rather than with countries that could pay for satellite communications. In his penultimate question, Welsh asked: "Since the government has borne the costs of development so far and will continue to bear substantial costs, is there any justification for considering private ownership at least until the major operational and international hurdles have been cleared?" These two themes would repeat themselves for months within the NASC: (1) U.S. prestige was more important than private profits, and (2) the government had paid for this development anyway.[41]

Over the next few days, Welsh spoke to or corresponded with most of the satellite communications players. On June 9 and June 12 Russell W. Hale (Welsh's deputy while he was hospitalized) and Welsh held meetings with staff members from NASA (Nunn), DoD (Ralph Clark), the State Department, the Office of Civil Defense Mobilization (OCDM), the Justice Department, and PSAC to discuss satellite communications. The outcome (on June 13, 1961) was the draft of a letter, from the president to the vice-president, asking the NASC to "make the necessary studies and government-wide policy recommendations for bringing into optimum use at the earliest practicable time operational communications satellites."[42]

On June 15 Kennedy signed the letter asking Johnson and the NASC to prepare a policy recommendation to accelerate the deployment of an operational communications satellite. Kennedy wanted the system to be global, sensitive to the needs of the developing world, and (he stressed) in the public interest. Welsh scheduled two more meetings of the interested parties, on June 27 and 28. The first meeting established the major problem areas: "(1) Foreign participation in the system and (2) ownership of the U.S. portion of the system." Both the FCC and DoD saw problems with any system in which the United States (specifically U.S. entities) did not own 51 percent or more. The Department of State thought it would be best "to bring foreign countries into [the U.S.] system, rather than have them decide to construct their own system." NASA pointed out that "pride and prestige" would be major factors for many nations, who would want "to use their own money and contractors to build satellites."[43]

The June 28 meeting had the same attendees but seems to have been dominated by the statements of NASA's Nunn. He pointed out that NASA was an R&D agency and would not attempt to build the entire system by itself. Furthermore, as Nunn stressed, the only active communications satellite launched to date (Courier) had a lifetime of only eighteen days. Obviously there was plenty of R&D to perform, especially in the areas of power and stabilization. AT&T had clearly expected to be the "chosen instrument" of U.S. satellite communications but had been rejected by the government. Nunn seems to have implied that NASA needed to examine all possible technologies but needed to let the private sector decide on the actual operational system. Nunn stated that any action by the NASC (it is not clear if he meant staff or the formal council) would not be final but rather that further executive branch decisions and guidance would be forthcoming.[44]

Ralph Clark, the DoD representative, brought up a concept that had been aired before: the combination of all U.S. international communications into a single coordinated entity. He then went on to suggest the DoD approach: LEO (low Earth orbit) and MEO (medium Earth orbit) first, followed by an Advent-like GEO system. He also pointed out that NASA and DoD were cooperating "across the board." Clark noted, "It would be difficult to expect private investment of any magnitude until some concept of the economic returns was available."[45]

Welsh continued to accumulate comments from the government, from industry, and from individuals. After some internal debate among the NASC staff, Welsh drafted a policy statement favoring private ownership and control. He circulated this draft among staff members from DoD, NASA, the State Department, the Atomic Energy Commission, the FCC, the Justice Department, OCDM, BoB, and the office of the science adviser. Before the paper was published, Welsh had publicly stated that the NASC did not favor government operation.[46] After presentation at the first formal NASC meeting on July 14, 1961, chaired by the vice-president, the statement was released by the president on July 24. The paper stressed that satellite communications was possible immediately and gave responsibility for operation to private ownership but gave responsibility for regulation, foreign negotiations, R&D, and launching to the government. It is interesting to note the changes between Welsh's early July draft and the final statement. Under ownership and operation, the first requirement was now "earliest practicable date."[47] Global coverage and foreign participation were moved from the last requirement to the second and third requirements. Under government responsibility, "use of the most effective techniques" had disappeared. After Kennedy's July policy statement, several liberal members of Congress sent him a letter suggesting that the government should avoid any decision that might result in a satellite communications monopoly.[48]

Congressional Hearings: July 13, 1961

The House Committee on Science and Astronautics communications satellites hearings, suspended in May, were resumed on July 13, 1961. In his opening remarks, Committee Chairman Brooks reminded the committee that the hearings had been suspended at the request of NASA, and he stated the purpose of the hearings:

> First, to acquaint the public with many of the details that have been worked out to make space communications systems an assured fact.
>
> No. 2, to lend what assistance this committee can give to NASA and to other agencies of the Government in furthering the great epochal achievement in communication; third, to determine the extent that private industry should participate in the space communication program, and fourth, to create a further sense of urgency among all involved in this important program.[49]

Brooks went on to suggest that a sense of urgency seemed to be lacking. He wanted to beat the Soviets and was concerned that similarities existed with the Vanguard-Redstone decision, which had allowed the Soviets to put a satellite in orbit before the United States. Congressman James Fulton (R, Penn.) challenged this view; to him, a "practical project" was more important than "shooting skyrockets."

Webb began by outlining events since January 1961. The first of these was the release on January 4, 1961, of a NASA request for proposals for the development of an experimental communications satellite. The second was a memorandum of understanding between the FCC and NASA. Webb's own words make his ideas quite clear: "We, in NASA, look to the FCC to take proper action on the problem of organizing the resources of private industry in such a manner as to meet governmental requirements and conform to public policy. On the other hand, we, in NASA, have the job of developing the space technology which any private organization authorized by the FCC will be able to utilize to provide communications services to the public." After his prepared statement, Webb was questioned by Brooks as to the financial contributions of private industry. Webb responded, "There are certain things no private industry can undertake on its own at this particular stage of the game."[50] This is a strange comment given the willingness of AT&T to fund communications satellite R&D by itself and given the "joint-venture corporations" formed by GE and Lockheed to develop and operate a commercial communications satellite system.

Webb, like most other government decision-makers, did not have a firm opinion about satellite communications. He seems to have been committed to private ownership of the satellite communications system as the "American Way"—preferably in the form of the FCC-proposed joint venture of the international communications carriers. On the other hand, Webb also seems to have been committed to NASA control of the space technology and space policy issues—this was NASA's charter. He was willing to launch an AT&T satellite in May, but he had not been willing in April, when his acrimonious correspondence with Kappel took place. This change was probably related to the increased importance that Kennedy's May 25, 1961, speech gave to the communications satellite program, but it may also have been an acknowledgment by Webb that his April reaction had been wrong.

Several congressional members had questions and comments. Congressman Fulton seemed to think that the FCC, NASA, and the NASC were delaying the establishment of an operational system. He was particularly annoyed by the number of economists on the NASC staff. Fulton felt that technical people would be more appropriate. Congressman William F. Ryan (D, N.Y.) wanted to know if government ownership was being considered for the operational system. Webb was simultaneously criticized for delaying AT&T and for allowing AT&T to gain an unfair (and monopolistic) advantage.

T. A. M. Craven represented the FCC at the hearings. Although Craven discussed the response to Docket 14024 and the FCC's decision, all the possibilities and problems of communications satellite policy were raised—none of them really having to do with the technology. The FCC felt it had the responsibility and authority to decide the issue. Congressman Ryan suggested that the Communications Act of 1934 might be in conflict with the Space Act of 1958, which put the responsibility for space activities in NASA. Craven indicated that the international carriers were willing to finance satellite communications. Congressman Ryan seemed to doubt this. Craven indicated that the NASC, if it overrode the FCC, would have to go to Congress for new legislation.

Committee Chairman Brooks had a specific agenda for satellite communications: "My interest has been largely in pushing a program so that we can show the American people a practical return for the vast sums of money which have been invested and which they are being called upon at this time and in the future to invest in the space program."[51] Brooks questioned Craven at some length on joint ventures, antitrust legislation, interference, and other matters. At one point he described the situation in Shreveport, Louisiana, when several competing organizations attempted to start television stations. The FCC had been forced to pick among them. With excessive delays, the FCC made a decision to provide television service through a pool until ownership questions could be resolved. Craven responded that a similar situation might occur with satellite communications and a similar solution might be proposed.

Craven attempted to explain away a comment, attributed to many different individuals, that satellite communications would be a $100 billion business by 1980. Brooks commented that he had heard that it would be a $250 billion business. Craven disagreed. The entire world telecommunications business might be worth that much, he said, but satellite communications would be only a small part of this and U.S. satellite communications an even smaller part. This led to a discussion of foreign participation in the satellite communications system. Ownership—both foreign and domestic—was apparently the largest issue, with significant fears of monopoly. Congressman James C. Corman (D, Calif.) was concerned that the billions of dollars to be spent would call for a huge investment by the U.S. taxpayer, who would insist that that investment be closely monitored—especially since government ownership was unlikely.

John H. Rubel, assistant secretary of defense and director of defense research and engineering, began the DoD presentation. He was accompanied by Ralph Clark, deputy director for communications. Rubel described the kinds of communications that were needed by DoD and the kinds of satellite systems that could satisfy those needs. He briefly described Advent and stated that geosynchronous systems were generally considered to be the most promising. West Ford, the "needles in space" passive reflector, was also briefly described. Encryption seemed to be a new term to some of the congressional members, who raised several issues concerning its need and cost. Rubel stated that the first Advent satellite would fly in an intermediate-altitude test orbit within a year.

Congressman Brooks asked if two systems, a military system and a commercial system, were really needed. He differentiated the Advent GEO system as one "not of great interest" to industry because of the technological problems involved (pointing, station-keeping, booster capability). Congressman Ken Hechler (D, W.V.) asked if Rubel had any cost-saving suggestions for the NASA-operated system. Rubel informed Hechler that DoD and NASA were working closely together on various technical problems. Congressman Corman brought up Edward R. Murrow's suggestion that the U.S. Information Agency should get a "special deal" from the satellite communications provider. Corman felt this was justified because of the taxpayers' investment in the system. Rubel was noncommittal. Fulton, as usual, asked some technical questions that seemed to roam around an issue rather than confront it. Finally Fulton seemed to become upset over the possibility that Cuba, Russia, and China might use the system. Rubel was followed by Philip Farley, of the State Department, who tried to explain that it takes two to communicate. Fulton did not like that statement either.

After adjourning on July 17, the committee did not reopen the hearings until August 9. In the interim, President Kennedy's July 24 "Statement on Communication Satellite Policy" was issued. Brigadier General Thames, commanding general of the AAMA, testified that there were no technical or organizational problems with the Advent project. He further testified that Advent could handle not just military but also government and commercial traffic. Four months earlier the director of defense research and engineering had considered terminating the program because of technical problems with both the launch vehicles and the satellite, as well as organizational problems that exacerbated the technical ones. In the months preceding the testimony by Thames, Rubel and Clark had been discussing with Seamans and Jaffe, of NASA, a joint NASA-DoD Hughes Syncom program, which would be in the best interests of the country—given the unfortunate status of Advent.

By mid-August all three civilian active satellite experimental programs—Relay, Telstar, and Syncom—would be officially approved. On July 27 NASA and AT&T entered into agreements for the reimbursable launch of Telstar. On August 10 Webb again testified before the committee. He described the AT&T agreements and allowed that there was one more system that looked "quite attractive." On August 11 NASA signed a sole-source contract with Hughes to build Syncom, the small geosynchronous satellite. In a 1960 conversation with Seamans, Glennan had suggested just such a "sole-source" procurement of the Hughes system after the general policy had been worked out.[52]

There was much confused discussion of the NASA program. Congressional members were pleased that AT&T, and not the taxpayer, was providing funding, but some had heard that AT&T was upset with what was happening.

AT&T had spent its own money and was now (possibly) to be left out in the cold. Congressman Ryan brought up the United Research report, which had recommended a government-operated system, at least initially. He told Webb that General Thames had testified that Advent could handle all the traffic and would be ready in 1965. Why were multiple systems necessary? Webb was followed by the NASC's Welsh, whose testimony seemed to further confuse matters. Both Webb and Welsh made it clear that only one system would be viable. According to Webb: "You simply cannot start two or three communication satellite systems. . . . Therefore, the Government policy has been to say we will create the conditions under which one system will be established."[53]

Two passive experiments were under way: Echo (NASA-AT&T) and West Ford (DoD). Two medium-altitude active experiments were under way: Relay (NASA-RCA) and Telstar (AT&T). Two geosynchronous experiments were under way: Advent (DoD-GE) and Syncom (NASA-Hughes). All the options were in play. Actual performance would determine the winner.

Points of View

By the fall of 1961, thanks to the congressional hearings and the presidential statements drafted by the NASC, it was possible to distinguish the various points of view of the industry and government players. It should be clear from the foregoing that there was no unanimous position within the federal government or, in most cases, even within a single department. NASA was apparently the most involved of the parties. AT&T was moving happily toward launching its satellite (Telstar, then called TSX) until Glennan informed the company, in late 1960, that launch services would not be available from NASA. Glennan seems to have favored AT&T's plans but felt that AT&T did not understand how to deal with politically sensitive issues. Webb seems to have been against the AT&T satellite plan, but he actually offered the company launch services in May 1961. Many civil servants within NASA seemed to feel that satellite communications was something the government should do and were looking forward to building and operating the final system. Nunn, assigned to develop communications satellite policy for NASA, seems to have felt that NASA's function was to perform R&D and allow industry to decide the form that satellite communications would take. All seemed to agree that NASA should "test" the technologies. Other issues, such as ownership and operation, were more contentious.

Many in Congress viewed the FCC as the appropriate policy forum. The FCC would prefer a system similar to, or evolved from, the transoceanic submarine cable arrangements. AT&T might control the U.S. portion of these cables, but the international record carriers (RCA, ITT, Western Union International) had rights of use and, eventually, rights of ownership. This suggested an international consortium led by AT&T but including other communications companies as well. This was apparently not a popular view either within the government or among the aerospace companies.

DoD had been cooperating with NASA but held a wide variety of views. Those at the very top were concerned that the coming "blackout" of HF radio communications due to the solar cycle might leave globally dispersed U.S. forces without real-time communications capability. Within the DOD satellite communications community were many who felt that Advent would fill both military and civilian needs. There were also concerns that Advent was already showing signs of failure. DoD often suggested combining cable and satellite entities into one national telecommunications entity—perhaps similar to Cable & Wireless in the United Kingdom. At this time DoD was convincing NASA that the Hughes geosynchronous satellite should be the beneficiary of the additional $50 million provided to NASA in Kennedy's May 25 speech.

The White House had similarly disparate views. The civil servants and academics seemed to favor a government system. Johnson may also have favored a government system. In one of several similar letters drafted by the NASC, Johnson responded to a query by stating, "The initiative and the funds to get started toward a communications satellite system have come mainly from the government."[54] On the other hand, all of the presidential statements seem to have allowed for private participation—even leadership—in the satellite communications endeavor.

The State Department—at least on the basis of the NASC records—was the department most interested in a specific outcome to the satellite communications policy process. Philip J. Farley, special assistant to the secretary of state for atomic energy and outer space, had argued—in the June 27 and 28 meetings and in numerous letters and telephone calls—that satellite communications was international by its very nature and that any ownership organization should be international in scope. He further argued that government ownership of the U.S. portion of an international satellite communications entity made more sense than ownership by private industry. His was almost the only voice that actively argued for a government-owned and -operated system—and was certainly the loudest such voice.

Congress voiced myriad views, many based on ignorance but many others addressing issues that the executive branch seemed to be ignoring. One group sincerely felt that the government should not be spending money on a commercial program when industry was ready to spend its own money. Another group argued that the government had made billion-dollar investments already. Some groups felt that satellite communications was either inherently governmental or inherently private. The various hearings and reports are illuminating, confusing, depressing, and entertaining—all at the same time.

Finally, industry also had mixed views. AT&T, the leading telecommunications company, thought that satellite communications was a multibillion-dollar business and would improve both costs and performance of transoceanic circuits. It was willing to fund the entire effort. In general, the other telecommunications companies favored an FCC-mandated consortium of some sort. This consortium would be similar to, but more "democratic" (i.e., less AT&T dominated) than, the existing submarine telephone cable arrangements. GTE was still talking to Hughes about a joint system. ITT was spending money on Earth-station design and was keeping some satellite R&D going. RCA seems to have wanted only to sell satellites to the government. Western Union International was not much of a player.

The aerospace (and electronics) companies wanted in on the deal—whatever it was. Both GE and Lockheed had volunteered to lead a consortium. Philco was still trying to sell to DoD (it would be successful). RCA simply wanted to sell products. North American Aviation was too busy with Apollo. Hughes wanted to start its own satellite communications company and had spent at least a million dollars of its own money on Syncom, but it preferred to allow the government to fund most early development—provided Hughes could keep its patent rights.

The Communications Satellite Act of 1962

In September 1961 Welsh brought Dr. Charles S. Sheldon on to his staff as a technical expert. Sheldon, the son of an engineer, earned his Ph.D. in economics from Harvard. He joined the Congressional Research Service (CRS), a branch of the Library of Congress, in 1955. When he joined the NASC staff, he was on extended leave from the CRS to serve on the House Space and Astronautics Committee. Several other staff members joined at the same time.[55] In October Nunn, of NASA, also became a formal part of the NASC team. The team represented a variety of viewpoints, but there was probably a general consensus that the government should control this technology.[56]

On October 11 the House Committee on Science and Astronautics issued a report based on the completed May through August hearings.[57] Of the eleven conclusions listed in the report, seven discussed the government's role in satellite communications and only four discussed the ownership and technology issues that were at the heart of the matter. The committee felt that the final system was part of the national space program and that as such, it should meet the nation's needs in the broadest sense, especially as a practical return on the dollars invested in space and as a manifestation of national prestige. No matter what the final ownership arrangements, the government must play the major role in R&D and in negotiations with foreign nations concerning operation of the system. NASA, DoD, and other government agencies must coordinate their actions in this area. The only comments on ownership were that it should not interfere with the early establishment of a global system and that it should not preclude serving unprofitable areas. The only comment on technology was a suggestion that NASA continue to explore all areas. As far as R&D and technological development were concerned, NASA had made a clean sweep, and industry was effectively excluded—except as NASA contractors.

Meanwhile, on October 13 the FCC Ad Hoc Carrier Committee recommended that the satellites be owned and operated by a nonprofit corporation formed by the international communications carriers. Both the Eisenhower and the Kennedy administrations had stated their commitments to private ownership. Congress had not spoken on the issue of ownership, but government control of the new technology was clearly important to Congress.

In November 1961 Kennedy asked the NASC to prepare a plan to implement the program defined in the July 24 statement. Welsh decided, possibly on his own, that implementation would require legislation. Welsh and Sheldon were the primary writers of this proposed (administration) legislation.[58] Welsh felt, as had Glennan before him, that there was a policy vacuum regarding communications satellites. Welsh was also concerned that if the proposed system was to be private, there must be competition. According to one NASC staffer, Welsh had strong feelings about AT&T's monopoly status. Sheldon wrote the first draft of the proposed legislation, but the draft was reviewed by all the parties on the NASC staff. One of these was William Meckling, who, along with Siegfried Reiger, was one of the prime analysts for the RAND studies of satellite communications. Farley, of the State Department, made further complaints that the final draft did not properly address international issues.[59] The final version was sent to the White House on November 30, 1961.

Congress had been studying the communications satellite issue since the beginning of the space age. As early as 1959 the House Committee on Science and Astronautics had held "Satellites for World Communications" hearings.[60] This interest persisted and was one of the forces that led NASA to look more closely at satellite communications in 1960.

In early 1962 three bills were introduced in the Senate: (1) the Kerr bill (S2650, on January 11), favoring ownership by the carriers; (2) the administration bill (S2814, on January 27), favoring broad-based private ownership; and (3) the Kefauver bill (S2890, on February 26), favoring government ownership.[61]

The Kerr bill was similar to the FCC position: ownership would be held by a consortium of existing communications carriers. Under the Kerr bill, a new corporation would be formed; 5,000 shares, at $100,000 each ($500 million capitalization), would be sold in minimum lots of five shares to U.S. common carriers authorized by the FCC. This would not allow AT&T to go its own way but would do little to minimize its domination of international telephony, since it would probably buy a plurality of the shares, if not a majority. Since Senator Kerr was a cosponsor of the administration bill, it is unclear how committed he was to his own bill. One author has suggested that Kerr might simply have been trying to make the administration bill look like the moderate choice between the carrier ownership of his bill and the government ownership of Kefauver's bill.[62]

The administration bill also called for a new corporation to be formed, but one with a broad base of ownership. Private citizens and other corporations could purchase shares. Ownership would not be limited to the international carriers. Limitations would be placed on the number of corporation shares that any single entity could own. Foreign participation would include ownership of shares as well as ownership of ground stations. Private ownership was assumed to maximize efficiency.[63]

The Kefauver bill was based on the senator's distrust of the monopolistic tendencies of industry—notably AT&T. This bill provided for a government-owned and -operated system. Senator Estes Kefauver (D, Tenn.) had three main arguments: (1) private ownership would constitute a monopoly; (2) an early system would need to be a low-altitude system, which would be inferior to the geosynchronous systems that would be delayed because of investment in the inferior system; and (3) satellites were being developed at government expense and the benefits should accrue to the public—not to profit-making private corporations. In hindsight it is obvious that the eventual Communications Satellite Act of 1962 did not delay geosynchronous systems: Comsat's first satellite, *Early Bird*, was geosynchronous. In addition, the costs of satellite development were borne by *both* the government and industry. AT&T had paid for development of the Telstar satellites and reimbursed NASA for the launches. Hughes had paid for development of the protoflight Syncom, although NASA had paid for the construction of the actual flight models. Only Relay was fully government-funded, and the remarkably short time between contract award and launch suggested that RCA had been spending its own money for some time. Only launch vehicles were government-funded—and need not have been. AT&T had considered building its own launch vehicles and had actually been approached by vendors. Wisely, AT&T had decided that this was one technology area where the government was likely to insist on a monopoly—its own.

Kefauver and his allies had little success with their three arguments. There was general agreement that a private corporation would be more efficient, but it was not clear that this efficiency would result in low prices to the public. Kefauver raised several foreign relations issues: international participation and service to the less-developed countries. Senator John Pastore (D, R.I.) defended the middle ground. He noted that the global system would be more efficiently operated by a private company but that government control must be maintained.

In February 1962 Dr. Donald R. MacQuivey, a Foreign Affairs officer from the State Department's Telecommunications Division, completed a staff report on communications satellites for the Senate Committee on Aeronautical and Space Sciences.[64] The report, almost three hundred pages long, was half discussion and half documentary.[65] MacQuivey provided an excellent summary of the issues and the various positions of industry, the FCC, NASA, the White House, and Congress over the previous few years. In his very first sentence he posed the basic question facing Congress and the government: "How can we get a workable system in operation as soon as possible?"[66] In his opening comments and his conclusions, MacQuivey stressed the duality of the problem. On the one hand, communications satellites were simply an "extension of the conventional means of communications." On the other hand, they were a major source of political prestige—both foreign and domestic—which created political problems in both development and operation. If private industry performed the communications satellite R&D, it would presumably be building claims to operate the final system. If the government performed the R&D, then it also built some claim to control of operations.

MacQuivey presented some interesting financial statistics. NASA funding for communications satellite R&D had grown from $3.1 million in FY60 to $29.5 million in FY61 and to $94.6 million in FY62 (including the March $10 million augmentation and the May $50 million augmentation). AT&T had a monopoly on the $42 million (1960) international telephone market, but the other carriers (principally RCA, ITT, Western Union) shared an $84 million international telegraph market. The telephone traffic growth rate was in excess of 20 percent each year. Telegraph traffic was growing at less than 10 percent each year. The projections for 1970 indicated that the total international communications market would exceed $1 billion. Large as these figures were, they were exceeded by the projections of Dr. Lloyd V. Berkner, chairman of the Space Science Board of the National Academy of Science, who was quoted as predicting a $100 billion market.[67] Whatever the size of the market, five of the nine carriers participating on the FCC Ad Hoc Carrier Committee expressed a willingness to put their own funds into a joint venture. These five were AT&T, ITT, Hawaiian Telephone, Radio Corporation of Puerto Rico, and Western Union. Among those unwilling to invest was RCA.[68]

Not least among those interested in the technology were the military services. MacQuivey pointed out that most military communications used HF radio. This service relied on the ability of the ionosphere to reflect radio waves in the 3–30 MHz band. Unfortunately, the ability of the ionosphere to reflect HF signals was dependent on the sunspot cycle. During sunspot minima, the ionosphere would not reliably reflect radio signals. During sunspot maxima, the solar flux might create noise that would interfere with communications. Satellite communications would thus be more reliable, especially in the event of a nuclear burst. In August 1958 a nuclear burst over Johnston Island in the Pacific disrupted communications for hours.

In the House, Congressman Oren Harris (D, Ark.) introduced H.R. 11040, identical to the Kerr bill, on April 2, 1962. On May 3 the bill passed the House, 354 to 9.[69] The Senate Committee on Aeronautical and Space Sciences reported favorably on Kerr's bill on April 2, 1962.[70] The Senate Commerce Committee also reported favorably on the Kerr bill, which was brought before the full Senate on June 14. Senator Kefauver and his allies attacked the bill for several days by means of an orchestrated filibuster. On June 21 the bill was withdrawn to allow other business to be completed. When debate on the bill resumed on July 26, a different climate prevailed: two weeks earlier, on July 10, AT&T had successfully launched *Telstar 1.* On August 1 the bill was referred to the Foreign Relations Committee, with instructions to refer the bill back by August 10.

Senators Wayne Morse (D, Oreg.), Albert Gore (D, Tenn.), and Russell B. Long (D, La.) opposed the bill in the Foreign Relations Committee, but most of their colleagues were eager to report the bill back to the full Senate. Both the secretary of state and the secretary of defense testified in support of the Kerr bill. In addition, the FCC—which had opposed the bill, feeling that only telecommunications carriers should be owners—now supported the bill. After approval by the Foreign Relations Committee on August 10, the Kerr bill was brought before the full Senate and was immediately the subject (again) of a filibuster by Senators Kefauver, Gore, Morse, Long, Ralph Yarborough (D, Tex.), Maurine Neuberger (D, Oreg.), Ernest Gruening (D, Alaska), and Paul H. Douglas (D, Ill.) On August 14 the Senate passed a historic cloture motion by a vote of 63 to 27. On August 17 the amended House bill passed the Senate by a vote of 66 to 11. On August 27 the amended bill passed the House, and on August 31 President Kennedy signed it into law.[71]

The 1962 Communications Satellite Act consisted of four major parts, or titles: Title I—Short Title, Declaration of Policy and Definitions; Title II—Federal Coordination, Planning, and Regulation; Title III—Creation of a Communications Satellite Corporation; and Title IV—Miscellaneous. Title I, Section 102, declared: "It is the policy of the United States to establish, in conjunction and in cooperation with other countries, as expeditiously as practicable, a commercial communications satellite system, as part of an improved global communications network." The emphasis within the act was on commercial and global communications, but use of the system for domestic communications was not precluded, nor was the creation of additional systems. Title II spelled out the responsibilities of the president, NASA, and the FCC.

Title III defined the new corporation. Sections 301 and 302 provided for the initial organization of the corporation. Section 303 defined the apportioning of the board of directors: three were to be appointed by the president, six were to be elected by stockholders who were communications common carriers, and six were to be elected by other stockholders. No communications carrier could vote for more than three candidates. Section 304 provided for the financing of the corporation. Communications common carriers would own 50 percent of the stock while the general public would own the other 50 percent. To encourage broad ownership, the act stated that the stock would be issued at a price of $100 or less. No noncarrier could own more than 10 percent of the stock. Section 305 defined the purposes and powers of the corporation.

Title IV, Miscellaneous, included Section 401, "Applicability of the Communications Act of 1934," Section 402, "Notice of Foreign Business Negotiation," Section 403, "Sanctions," and Section 404, "Reports to the Congress."

Sputnik 1 was indeed a spur to concrete action, as John Pierce had suggested. In 1958 AT&T and the Jet Propulsion Laboratory piggybacked a communications experiment (Echo) on a NACA atmospheric probe. In 1959 AT&T and Hughes began to design active communications satellites using their own funding. In 1960 AT&T applied to the FCC for permission to launch an experimental active MEO communications satellite. In 1960 Hughes actually built a prototype of its active GEO commercial communications satellite. By 1960 most communications and electronics companies were pursuing some level of communications satellite R&D. This increased attention caused the government (by mid-1960) to perceive the need for a communications satellite policy. NASA grasped control of satellite technology development by controlling the allocation of launch vehicles. The FCC was ready to treat satellite communications as simply another communications technology. This was not enough for many congressional members and civil servants who, for a variety of reasons, felt that the government should control this technology. Major Gagarin's flight and the Bay of Pigs provided a foreign prestige rationale for visibly beating the Soviets in this new space application. In the summer of 1960, industry was in control of satellite communications. By the summer of 1962, the government controlled satellite communications. R&D would be controlled by NASA. Operations would be controlled by the public-private entity formed by the Communications Satellite Act of 1962.

5. Building the Satellites

The answer we want is full speed ahead. —Representative Joseph E. Karth (D, Minn.), August 1962

During 1961 and 1962, three communications satellite designs were under construction: the RCA MEO (medium Earth orbit) satellite, the AT&T MEO satellite, and the Hughes GEO (geosynchronous Earth orbit) satellite. Within two years, all three satellite designs would be successfully launched.

Hughes Aircraft Company

Hughes had begun redesigning the Syncom satellite to accommodate launch on the Thor-Delta from Cape Canaveral rather than on the Scout from Jarvis Island. Several new technical challenges had to be met. The launch vehicle would place the spacecraft into an elliptical geosynchronous transfer orbit (GTO). Circularization of this highly eccentric orbit would require the use of an apogee kick motor (AKM), which would be controlled by the spacecraft. The Delta had a "tip-off" error associated with separation of the satellite-AKM package from the launch vehicle. This would cause an attitude error that might affect the efficiency of the AKM thrust vector. However, the added capability of the Delta meant that Syncom mass could grow from 30 pounds to 60 pounds.

The added weight might also allow for added design features. Even greater

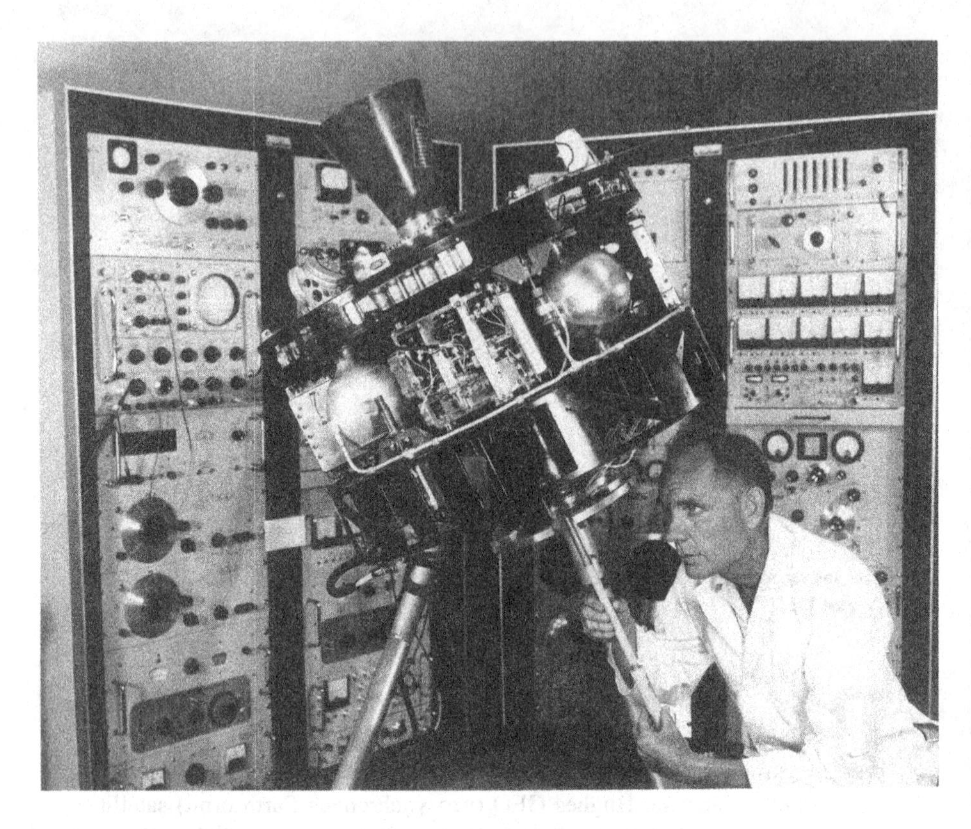

Tom Hudspeth testing Syncom in 1962. (Courtesy Hughes/BSS)

weight would allow the Syncom design to accomplish all the myriad goals of the satellite communications studies. By the fall of 1961, Harold Rosen of Hughes and Leonard Jaffe of NASA had begun discussions of potential improvements to Syncom. In February 1962 Hughes presented its plans to NASA. In March Hughes prepared a brochure describing the Syncom *Mark II* satellite.[1] *Mark II* would be launched on the Atlas-Agena and would weigh approximately 1,000 pounds including AKM. The solar panels would generate over 100 watts, compared with the 25 watts on the Syncom *Mark I.* Four complete communications repeaters would be simultaneously available for communications. The effective isotropic radiated power (EIRP) transmitted to the ground by each of the four transmitters would be many times greater, due to the higher gain of a phased-array antenna, which would concentrate the RF (radio frequency) energy on Earth rather than radiating most of it into space. *Mark II* would also be reliable. The predicted lifetime was five years.

At about this time, Hughes senior management began to plan the future of Syncom. The Advent troubles had helped sell the program in the first place; perhaps Syncom could replace Advent. AT&T had seen the direction that Congress was taking and was losing interest in medium-altitude systems because it seemed unlikely that AT&T would be a major factor in the eventual operating company. The Hughes Syncom could be sold to the congressionally approved operating entity, to the U.S. military, to foreign customers, and perhaps to NASA as an extension of its communications network. In addition, a dedicated live intercontinental television satellite was a real possibility. All of this required that Hughes protect its patent rights.[2]

AT&T

Although Hughes would spend three decades fighting for its patent rights, AT&T was hit with a variety of restrictions before it was allowed to participate in the communications satellite program. NASA would have the rights, including licensing rights, to all AT&T satellite communications inventions after May 1961. NASA would conduct all of the negotiating with foreign post, telegraph, and telephone administrations (PTTs). NASA would coordinate all tests. NASA would handle all publicity. None of these restrictions were acceptable to AT&T, but it was more interested in building satellites and ground stations than debating arrangements.

By January 1962 the AT&T Andover Earth station was essentially complete and ready for testing. The first satellites—*TSX-1A* and *TSX-1B*—were also ready for testing. *TSX-1A* was scheduled for an April 1962 launch and the backup *TSX-1B* for an October 1962 launch. *TSX-2* was identical to *TSX-1,* but *TSX-3* and

TSX-4 would be much heavier satellites launched by Atlas-Agena into 6,000-mile circular, polar orbits—nominally above the lowest van Allen radiation belt. The *TSX-3* specification was somewhat indeterminate, but *TSX-4* was intended as a prototype for the operational system. A late 1963 launch was envisioned for *TSX-4.*

AT&T's 1962 budget for communications satellites was $18 million, of which $3.5 million was allocated to finish Andover and $5.3 million to support other ground-station work. A. C. Dickieson, a senior Telstar official, recommended that instead of building *TSX-2,* AT&T should launch the *TSX-1B* satellite with whatever modifications could be made cheaply, and that more funds should be devoted to *TSX-3.* This more complex satellite would have a controlled attitude—probably using gravity-gradient techniques—and a more extensive communications payload.

By February 1962, with several communications satellite bills moving through Congress, it was clear that AT&T's investment in satellite manufacturing would be wasted. The operational system would be a government-sponsored private enterprise or simply government-operated. There would be no place for AT&T satellites. AT&T had built an expensive ground station. With many satellites in low orbit, it made sense to build simple satellites and complex ground stations. The horn-and-MASER AT&T ground-station design would lose out to the cheaper parabolic dish-and-parametric amplifier designs like ITT's. There would be no profit from Telstar, but there was still pride in performance. Problems discovered in testing caused AT&T to request a launch delay. NASA took advantage of this opportunity to change some of the contract agreements. AT&T's pride was taking a lot of hits.[3]

NASA

On March 31, 1962, a NASA Communications Program Review was held. In presentations to NASA senior management, managers from the NASA Goddard Space Flight Center and NASA Headquarters described past accomplishments and future plans. Morton J. Stoller, NASA director of applications, opened by stating, "The active satellite program grew from nothing to three flight projects in the first nine months of 1961."[4] Leonard Jaffe followed Stoller and listed the three major objectives of the NASA communications satellite program: (1) demonstration of feasibility, (2) establishment of operational systems, and (3) continued R&D support. This program would help to determine the best system (active/passive, low/high), but the remaining problem—who would operate the system—needed to be solved elsewhere.[5]

Jaffe briefly summarized the technological options and emphasized the uncertainties regarding the near-term technical feasibility and economics of all but the simplest systems. One example given by Jaffe was the disagreement regarding the appropriate modulation technique. FM (frequency modulation) minimized required satellite power but was wasteful of spectrum. SSB (single sideband) minimized spectrum use but required very high-power satellites. PCM (pulse code modulation) was intermediate between FM and SSB. There was also the problem of multiple access. An operational system would have to provide the capability for more than just two Earth stations to communicate. Jaffe, like many others, looked forward to the day when Saturn-class launch vehicles would be able to put nuclear-powered satellites into orbit—probably in the post-1968 period. Jaffe's vision was somewhat reminiscent of the very large manned communications satellites envisioned by Arthur C. Clarke and others.[6] Jaffe also mentioned the activities of the CCIR (International Consultative Committee on Radio) Study Group IV, which had met in Washington in February. The consensus of the meeting was that communications satellites would share the ground microwave frequencies (e.g., 6 and 4 GHz). Telstar already used these frequencies for both uplink and downlink.

At this time NASA was still studying passive satellite technology. The success of *Echo 1* was coupled with a realization of the constraints imposed by passive satellites, but NASA planned a more capable follow-on in *Echo 2* and a multiple launch system in the Rebound program. *Echo 2* was eventually launched, as was the Department of Defense (DoD) *West Ford* "needles" constellation, but passive satellites were an idea whose time had passed.

The Relay project presentation was prepared by the project manager, Joseph Berliner, but was delivered by Lieutenant Colonel R. E. Warren, from Jaffe's communications group at NASA Headquarters.[7] The main objectives of the program were (1) demonstrating the feasibility of LEO (low Earth orbit)—actually medium-altitude—active communications satellites and (2) examining the effects of radiation on spacecraft electronics and solar cells. Because of the limited Delta launch vehicle performance, the Relay orbit was planned to be highly eccentric—traversing the lower van Allen belt as it traveled from apogee to perigee. The orbit, inclined at 54 degrees, would have an apogee of approximately 6,000 kilometers and a perigee of approximately 1,500 kilometers.

The 150-pound (70-kilogram) spacecraft had an unusually light structure (less than 10 percent of total weight) and high-power RF output due to the advanced RCA 10-watt TWT. A series of design changes had increased the weight of the spacecraft from 93 pounds (42 kilograms) to approximately 150 pounds. The mass breakdown by subsystem is shown in Table 5.1.

These changes included spin stabilization, wideband redundancy, active thermal control, and two-way telephony. The spin-stabilization system included a

Table 5.1
Relay Mass by Subsystem

Subsystem	Mass (pounds)
Communications	32.4
TT&C	16.3
Structure	14.1
Thermal Control	3.0
Power	33.6
Radiation Experiment	13.2
Harness	7.4
Solar Panels	25.8
Attitude Control	1.5
Total	147.3

coil to create precession torques, a horizon scanner, and a sun sensor. The communications frequencies were changed so that at least the downlink (4,170 MHz) was the same as Telstar. The presentation emphasized the "superior" communications capability of the Relay spacecraft compared with that of the Telstar spacecraft. The ground stations for Relay were essentially the same as those for Telstar. The Andover, Maine, and the Pleumeur Bodou, France, stations were 60-foot (18-meter) horn antennas designed by AT&T. The British Post Office (BPO) antenna was an 85-foot (27-meter) parabolic dish. Nutley, New Jersey, and Rio de Janeiro, Brazil, would have 40-foot (12-meter) and 30-foot (9-meter) dishes designed by ITT. The ITT designs were suitable for telephony. The AT&T and BPO designs included television testing provisions.

Charles P. Smith, project manager for the NASA Goddard Space Flight Center, prepared the Telstar presentation, which was also delivered by Warren.[8] Telstar was planned for launch into a 500-by-3,000-nautical-mile orbit (1,000-by-6,000 kilometers), inclined 45 degrees. Special mention was made of the fact that Relay weighed only 147 pounds, compared with Telstar's 175 pounds—thus allowing a slightly higher, more inclined orbit for Relay. Although generally similar to Relay, Telstar had many specific differences. Perhaps the biggest difference was the open structure of Relay. By contrast, all of the Telstar electronic subsystems were enclosed in a foam-filled canister that was welded shut and attached to the spacecraft shell with nylon lacing.

The Telstar uplink was at 6,390 MHz and the downlink was at 4,170 MHz. These were the land microwave frequencies that AT&T had originally described, in the "above 890" decision, as unsuitable for satellite communications. Communications experiments would include a variety of voice, data, and television tests. The AT&T Andover Earth station and its twin in France had MASER receivers that, when combined with the high-gain low-noise characteristics of the 60-foot horn antennas, provided a unique combination of very high gain and very low noise.

Warren also delivered the Syncom presentation, which had been prepared by Alton E. Jones, the project manager.[9] The Syncom presentation began with a short review of the program. According to Jones, NASA had decided to pursue the January 1960 Hughes unsolicited proposal after making a few changes: (1) the communications capability would be narrowband, conceivably as little as one two-way voice channel, (2) the satellite would be launched on a Delta rather than a Scout launch vehicle, and (3) the launch site would be Cape Canaveral, Florida, rather than Jarvis Island in the Pacific. No mention was made of the $50 million communications satellite supplemental in mid-1961. Jones pointed out that Syncom was a joint NASA-DoD program, with NASA providing the space segment and DoD providing the ground segment. The Syncom project had the following five objectives:

1. Provide twenty-four-hour orbit experience
2. Flight-test a simple approach to orbit and attitude control
3. Develop capability of achieving GEO using an apogee kick motor (AKM)
4. Test component life at GEO
5. Develop transportable ground facilities

The total Syncom weight had increased dramatically. About half the total would be allocated to the solid AKM (compare Table 5.2 with Table 3.1).

One of Syncom's unique features was its attitude- and orbit-control capability. The propulsion system of AKM, hydrogen peroxide thrusters, and nitrogen thrusters provided the Δv (velocity changes for controlling the orbit) and torques (for controlling attitude), but the system that controlled the Δv and torques was extremely simple and extremely clever. The spacecraft was attached to the launch vehicle at the top or north end of the spacecraft. The AKM was attached to the other end so that at launch, the AKM was the topmost item in the "stack." This was done so that the AKM would be facing in the proper direction after the spacecraft reached apogee. At that point the AKM would be fired by command or by a timer started by the spacecraft launch vehicle separation switch. A spin rate of 160 RPM provided stability during the firing. Exact GEO velocity was obtained by firing the hydrogen peroxide and nitrogen

Table 5.2
Syncom Mass by Subsystem

Subsystem	Mass (pounds)
Electronics	16.3
Harness	1.6
Power	8.2
Attitude Control	10.1
Propulsion	21.1
Structure	15.1
AKM	52.1
Total	124.5

thrusters. After GEO was achieved, an axial thruster parallel to the spin axis was pulsed to rotate (precess) the spacecraft into the orbit normal position so that the spin axis was aligned with Earth's north-south axis. This pulsing was achieved by firing the thrusters a specific time after the sun sensor detected the sun. Radial thrusters firing perpendicular to the spin axis could be used to adjust the spacecraft orbit over time. There were many other unique features on this lightweight spacecraft, but this attitude- and orbit-control system was the breakthrough that made GEO communications satellites possible.

Although the Syncom communications capability was often disparaged as "only one voice channel," it had one big advantage over the Relay and Telstar systems: the Syncom transponder could be used continuously—not just when the batteries were fully charged and not just when it was temporarily in view. The nominal 25 watts generated by the solar panels could support both bus and payload. The Hughes TWT had achieved its objectives of lightweight and low-input power (i.e., high efficiency).

The tracking, telemetry, and control (TT&C) ground stations would have 30-foot (9-meter) diameter dish antennas. One transportable TT&C system would be located at Lakehurst, New Jersey. The other, mounted on a converted Victory ship, would be on the West coast of Africa. The communications ground stations would be the Advent stations at Camp Parks, California, and Fort Dix, New Jersey.

Jaffe prepared and presented the Advanced Systems presentation.[10] Studies had been under way for some time regarding Relay and Syncom follow-on systems. The Delta launch vehicle capability had severely limited what was achievable. Future systems would use the Atlas-Agena launch vehicle, which had significantly greater capability. Instead of putting 125 pounds (57 kilograms) in a highly eccentric 900-by-3,000-mile (1,700-by-5,500-kilometer) orbit, the Atlas-Agena could put 600 pounds (270 kilograms) into a circular orbit at 12,000 miles (with a 12-hour period). The extra mass would allow for inclusion of four television channels, multiple-access capability, and gravity-gradient stabilization. Ultimately, nuclear power might be the power system of choice. Similarly, 500 pounds could be placed in geosynchronous equatorial (geostationary) orbit. This satellite would also have four television channels but would still be spin-stabilized. Increased gain would be achieved using a phased-array antenna.

Jaffe also made the International Committee, Demonstration Committee, and Ground Station Committee presentations, which had been prepared by Arnold Frutkin, Harold Goodwin, and Jaffe respectively.[11] These summarized the efforts to develop a test plan using U.S., British, French, German, Italian, and Brazilian Earth stations. Jaffe pointed out that these Earth stations represented investments of several million dollars by each country involved. The public relations aspect was not forgotten either. Jaffe emphasized that the Ground Station Committee attempted to convey the impression of a "single, unified United States cooperative program."

Morton J. Stoller, Jaffe's boss at NASA Headquarters, made the concluding summation.[12] Some of Stoller's comments, especially on the use of nuclear power and arc-jets, were in the nature of science fiction, although this may not have been evident at the time. Stoller did suggest that work on the twenty-four-hour communications satellite should be coordinated with work on the twenty-four-hour meteorological satellite—prefiguring the later ATS program. He also suggested that NASA should work on communications satellite ground systems. Other discussion issues included transmission of the 1964 Tokyo Olympic games and staffing at the NASA Goddard Space Flight Center.

During the comment session after Stoller's remarks, Harry Goett, director of the NASA Goddard Space Flight Center, discussed the implications of the "chosen instrument" (Communications Satellite, or Comsat, Corporation) that was expected to result from the congressional bills recently introduced. He was especially interested in performing R&D on the operational system that might be required within six months. Goett specifically commented on AT&T's unwillingness to deal with foreign stations through NASA.

Dr. Hugh Dryden, the deputy administrator of NASA, was uncomfortable with the rush to advanced programs. He commented, "You can develop something in a Government laboratory [but] you'll have a great deal of trouble getting that into use by a commercial group." He was also unconvinced that NASA should perform ground-station R&D. He stated, "This certainly is not the intent of Congress." Dryden continued in this vein, advising Goett to "think through" the areas that he might want "to get into."

The ambitious NASA communications satellite program described in the March 31, 1962, program review was immediately in the news. The communications satellite budget was reported to be $85 million in fiscal year 1963 (FY63) and even higher in FY64. Stoller was quoted as saying, "When all the technology is in hand, it will be the synchronous-orbit satellites which will be the most attractive to us."[13] The Hughes effort with Syncom and Advanced Syncom (the Syncom *Mark II*) was very much in the news. The June 1962 issue of *Interavia* had several articles on communications satellites. One of these, by Samuel G. Lutz of Hughes, explained how GEO communications satellites would not interfere with terrestrial use of the common-carrier frequencies (6 GHz and 4 GHz)—a benefit unique to GEO.[14] On June 18, 1962, NASA formally announced that a $2.5 million contract had been awarded to Hughes to study problems associated with the 500-pound, Atlas-Agena-launched Advanced Syncom.[15]

Artist's concept of Advent satellite. (Courtesy Black)

Table 5.3
Advent Budget Status: End of FY62

Item	Contractor	Expenditures	Salvage Value
Atlas-Agena (2)	Convair	$15.4 million	$12.8 million
Satellite (s)	GE/Bendix	$74.8 million	$15.1 million
TT&C Stations (4)	Philco	$12.5 million	$12.0 million
Communications Stations (3)	Sylvania	$54.1 million	$40.3 million
Total		$156.8 million	$80.2 million

Source: U.S. Congress, House, Committee on Science and Astronautics, *Hearings: Project Advent-Military Communications Satellite Program,* 87th Cong., 2d sess., 1962, 30.

Advent

DoD had been planning communications satellite programs since before *Sputnik 1.* Limited success had been obtained with Score and Courier, but since 1959–60, DoD had placed most of its resources in the Advent geosynchronous communications satellite program. Problems had risen with the GE-built satellite, the Atlas-Centaur launch vehicle, and the program management organization. Major committee studies and reports had been presented in May 1961, December 1961, and May 1962. This last report, by Ralph Clark and based on April 1962 studies by the director of defense research and engineering (DDR&E), recommended the cancellation of the geosynchronous Advent program and a reorientation toward smaller, simpler, medium-altitude systems.

Costs had escalated during the life of the program. In February 1960 the Advanced Research Projects Agency (ARPA) estimated that the entire program would cost about $159 million. The September 1961 estimate was $352 million. The General Electric (GE) spacecraft had originally been estimated to cost $56.5 million. The DDR&E estimated that costs would have risen to $180 million without redirection. As of the end of FY62, actual expenditures for Advent were as shown in Table 5.3.[16]

The Army recommended changes in the program in February 1962—essentially canceling the later phases of Advent. In April the DDR&E completed a study—ultimately the May 22, 1962, Clark paper—and on May 23, 1962, the secretary of defense essentially recommended canceling the Advent program and initiating a lightweight satellite program. Congress was not pleased to hear that a multimillion-dollar program was being canceled as a failure. On June 25, 1962, George P. Miller (D, Calif.), the new science committee chairman, wrote Dr. Harold Brown, the director of defense research and engineering, and asked a series of questions concerning the demise of Advent. On July 10 Brown responded. Hearings were held on August 15 and 17.[17]

The hearings demonstrated the lack of technical understanding within Congress as well as a concern that there was a problem that needed fixing. Much of the overrun was due to excessive use of overtime in an attempt to meet schedules and to inefficiency. DoD was still convinced that GEO was where it wanted to be, but it was also convinced that the technology for GEO was simply not ripe. It was therefore committed to a medium-altitude satellite to provide an interim capability. The interim system would provide communications during the sunspot minimum of 1964, when some HF frequencies would "disappear." The GEO system would be available during the 1969 sunspot maximum, when some HF frequencies would become extremely noisy. Representative Joseph E. Karth (D, Minn.) remarked toward the end of the hearings, "The answer we want is full speed ahead on Syncom or its counterpart in the Department of Defense, and use that system as soon as it is available."[18]

Telstar 1 Launch: July 1962

Less than a year after the July 27, 1961, AT&T-NASA agreement, AT&T was ready to launch its *Telstar 1* satellite. The Andover station, with its huge $10 million 60-by-60-foot horn antenna and MASER-amplifier receiver, had been ready since the beginning of the year and was officially operational in April.[19] The Pleumeur Bodou station in France, essentially a copy of the huge AT&T horn, was started in February and finished on July 7, 1962. Three days of tracking and RF calibration ensued between that date and the launch. The British station at Goonhilly was also ready for launch.

Part of the preparations for launch were the arrangements made for the media and public relations personnel. Somewhat to the surprise of the technical managers, James Dingman, executive vice-president of AT&T, had scheduled an extravaganza for the first Andover pass. The managers' concern was that tracking the satellite was in many ways the biggest unknown. They were convinced that if the Delta put a functioning Telstar in orbit, everything else would work perfectly—except tracking, which could not be tested before the launch. Dingman's rationale was that this might be an opportunity to provide senators, who were debating the Communications Satellite Act, to see the benefits of private industrial expertise. NASA, of course, objected.[20]

The Andover ground station began its formal pre-launch activities at 8:00 A.M. EDT on July 9, the day before launch. Various minor problems were encountered, including the shorting of a klystron power supply. A backup was started on its way from Murray Hill, New Jersey, by car (probably a twelve-hour-plus

Telstar 1 launched on Thor-Delta, July 10, 1962. (Courtesy NASA)

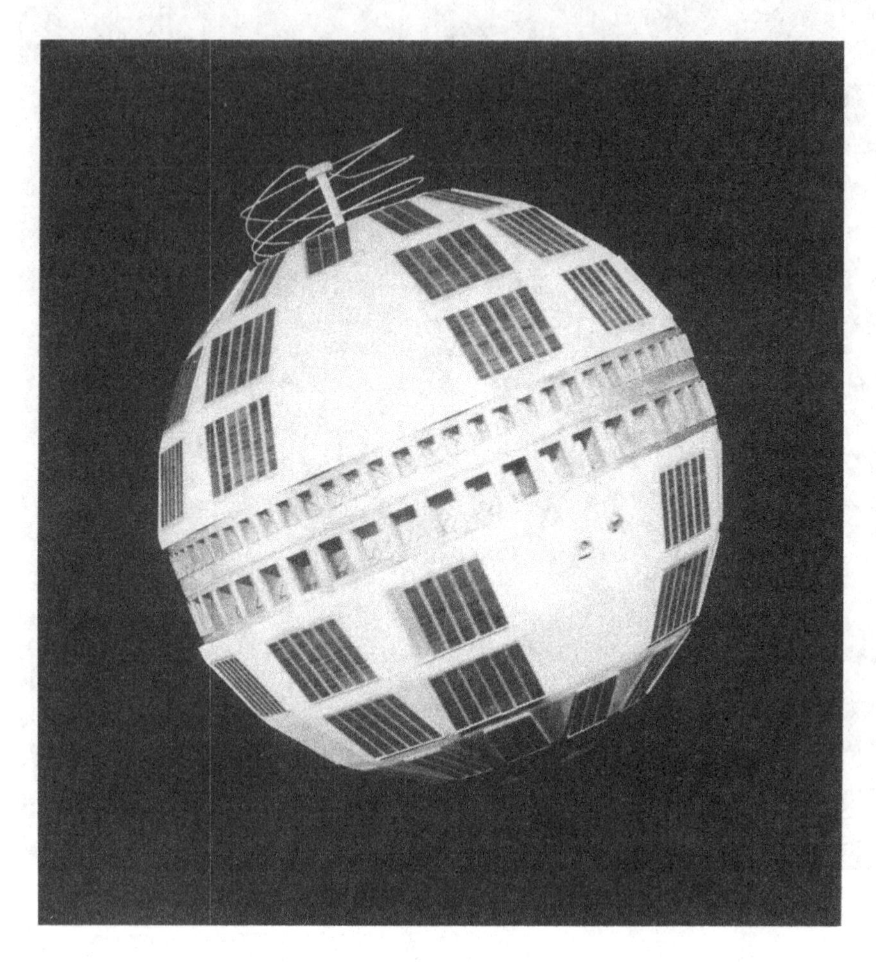

Telstar satellite. (Courtesy NASA)

L. S. Miller (AT&T vice-president) and E. F. O'Neill (director of satellite communications for Bell Telephone Laboratories) donating the Telstar flight model to the Smithsonian, July 10, 1963. (Courtesy AT&T)

drive in 1962). At 4:00 A.M. EDT on July 10 Cape Canaveral began its terminal count. At 4:25 A.M. EDT the Delta, carrying *Telstar 1,* lifted off from its pad at Cape Canaveral. At 2:36 P.M. EDT Pleumeur Bodou technicians reported they had been tracking the 136-MHz VHF telemetry beacon since 2:02 P.M. EDT. At 4:45 P.M. EDT Andover was able to track the 136-MHz beacon for several minutes, allowing AT&T engineers to examine telemetry data, which indicated that all was well.[21]

At 7:18 P.M. EDT the first pass with mutual visibility (both stations could see the satellite) between Andover and Europe began. After telemetry indicated all was well, the command to turn on the communications repeater was sent. At 7:25 P.M. EDT the Andover transmitter was turned on. At 7:47 P.M. EDT the Pleumeur Bodou station reported receiving an excellent video signal from Andover. Goonhilly was unable to receive a good signal due to ground-station problems. On the next pass, at 8:19 P.M. EDT Andover was able to view the signal received at Holmdel, New Jersey, by microwave link. At 8:26 P.M EDT telephony signals were transmitted. In its first twenty-four hours of operation, *Telstar 1* had demonstrated transatlantic television and telephony. During the 7:30 P.M. EDT pass, the public relations extravaganza had gone into effect, including a phone call from AT&T President Fred Kappel to U.S. Vice-President Lyndon Johnson and a videotape of the U.S. flag waving in front of the Andover radome. The engineers thought this was corny and were surprised to hear later that this had brought tears to the eyes of their friends and neighbors. During the telephony testing during the next pass, Jim Fisk, head of Bell Telephone Laboratories (BTL), got the first wrong number by satellite when attempting to call his wife. On the following day, the British, having corrected a misunderstanding on the sense of the circular polarization of the satellite signal, were able to participate. Both Goonhilly and Pleumeur Bodou transmitted television to Andover successfully. Satellite communications was a reality—and AT&T had been first![22]

The Delta launch vehicle placed *Telstar 1* in a 529-by-3,531-statute-mile orbit inclined 44.8 degrees to the equator. Immediately after launch, *Telstar 1* was in a terminator orbit such that it was always in sunlight. As the relative aspect of the sun and the orbit plane changed with time, *Telstar 1* was subject to daily, thirty-minute eclipses for most of the year. The initial spin rate was 178 RPM but this had been reduced to 146 RPM by mid-September. The magnetic torquing coil was available to precess the satellite's attitude to improve the solar aspect angle, but this proved not to be necessary. Solar array current had a noticeable 0.5 amp ripple due to the uneven distribution of solar cells. The average output was about 5 amps at launch and about 4.5 amps by September. Turning on the TWT caused the batteries to discharge—the solar array generated insufficient power to support the communications payload. Observation of the shape of the battery charge-discharge curve suggested that no battery deterioration had occurred. Solar cell degradation was consistent with pre-launch estimates and indicated that after two years, the output of the solar array would be about 68 percent of the beginning-of-life (BOL) output. Thermal control remained nominal during this period. All in all, the *Telstar 1* satellite was working as expected.[23]

During the first few days in orbit, various qualitative tests were made showing that *Telstar 1* was adequate for television, telephony, and data transmission. During the following months, more precise tests were performed. Noisy signals prevailed at ground antenna elevations below about 4 degrees. Noise temperature was observed to vary with elevation and rainfall. In addition to the BTL MASER receiver, a liquid-nitrogen-cooled parametric amplifier was tested at the ground station; the MASER had a 4-dB lower noise figure (i.e., it was better than expected). During this two-month period the sun was observed directly

Alton C. Dickieson, Telstar program manager. (Courtesy AT&T)

behind the satellite, causing loss of signal (sun outage). As little as 1 degree in separation (the sun is 0.5 degrees wide, and the Andover antenna had a beam width of a little less than this) avoided this problem. Gain-frequency response was measured and showed a tilt attributed to ground equipment rather than the satellite. Other performance measurements were made to evaluate the ability to transmit television, telephony, and data. All showed excellent performance.[24]

The *Telstar 1* radiation environment experiments were well thought out—NASA had chosen the AT&T experiment design for Relay. The van Allen radiation belts were a concern to designers, although the belts were not considered a major barrier to successful satellite communications. The radiation experiments covered two broad areas: damage to electronics (solar cells and transistors) and actual measurement of energetic particle fluxes (electrons and protons). AT&T had chosen n-on-p solar cells based on laboratory experiments. Other experiments had shown the relative degree of damage to be expected from electron and protons of different energies. The solar array degradation suggested 1 Mev electron fluxes of about 6×10^{12} electrons/cm^2/day or 10 Mev proton fluxes of about 10^9/cm^2/day. In addition to the solar cells in the power-generating array, a series of test cells were mounted on the satellite. These had covers of 20-, 25-, or 30-mil-thick sapphire. In addition, a previously irradiated cell was mounted on the spacecraft. After a little more than a month, the 20-mil-cover cells showed a 20 percent degradation while the 25- and 30-mil-cover cells showed a degradation of only 15 percent. Similarly, transistors were shielded with 3-mil-Kovar, 3-mil-Kovar plus 11-mil-aluminum, and 3-mil-Kovar plus 35-mil-aluminum. Not surprisingly, the heavier aluminum shielding fared better. All fared poorly. *Telstar 1* experimenters estimated that one-third of the damage was due to electron flux and two-thirds to proton flux. There was some speculation that the highest-altitude electron flux was caused by the July 9, 1962, explosion of a nuclear weapon in space (*Starfish*).[25]

In its first two months in space, in spite of a higher radiation environment than expected, *Telstar 1* proved that medium-altitude communications satellites were viable. In testimony before Congress on October 4, 1962, A. C. Dickieson, a thirty-nine-year AT&T employee and the executive director of the Transmission Systems Division at Bell Labs, presented AT&T's conclusions from the *Telstar* experiment:

1. Telstar signals penetrate a short atmospheric column minimizing losses.
2. Heavy rain increases noise, but less than expected.
3. Telstar radio equipment worked in space exactly as it did in the laboratory.
4. The inner canister maintained temperatures between 60–75° F.
5. Spin stabilization is successful, but magnetics cause the satellite to spin down.

6. Radiation damage to the solar cells will reduce power by about one-third in two years. Damage to transistors seems to be minimal—possibly because of the conservative design.
7. Concerns about tracking a small satellite with a "pencil beam" appear to have been groundless—there had been no problems.[26]

Telstar 1 was a roaring success. It had demonstrated television, telephone, and data links between the AT&T stations in the United States and stations in the United Kingdom and France. Most of all, *Telstar 1* had captured the imagination of the world.

Congress and Comsat

In late September and early October 1962 the applications subcommittee of the House Committee on Science and Astronautics held the "Commercial Communications Satellites" hearings. Ken Hechler (D, W.V.) opened the hearings by stating that their major purpose was to determine "the most effective and least expensive system for commercial development."[27] Specifically, should the system be medium-altitude or geosynchronous? Witnesses at the hearings included representatives of Hughes, NASA, DDR&E, the State Department, the U.S. Information Agency (USIA), and AT&T. By far the longest presentations were those of Hughes and NASA. Hughes led off the presentations.

On September 18, 1962, Fred Adler, the Space Systems Division manager, and Gordon Murphy (another CalTech graduate), the Syncom program manager, made the case for geosynchronous satellites of Hughes design. Murphy appears to have startled the members of Congress by stating that in addition to the NASA contract for three *Syncom I* satellites, a NASA study contract for advanced satellites was also in place. He added: "We expect that the later contract will lead to an initial operational communications satellite demonstration in the first half of 1964. We call the advanced satellite Syncom *Mark II*." Murphy pointed out that from 1959 to 1961, Hughes had been working on its satellite design and had even constructed a complete satellite for environmental testing. All this had been done using company funds. The result was that Hughes would launch its first Syncom satellite only seventeen months after signing a contract. This was less than RCA's nineteen months from contract to launch but more than AT&T's fourteen months from contract to launch. Murphy told the committee that what was needed was a NASA commitment by March 1963 to build flight vehicles and a NASA commitment to build ground stations. The Telstar, Relay, and Syncom ground stations were all owned by others. NASA was building a station in Rosman, North Carolina, and the Jet Propulsion Laboratory had a station in the Mojave Desert near Goldstone, California. Murphy argued, "A synchronous system can be installed sooner than a medium-altitude system because fewer satellites will have to be launched and because the ground stations of a stationary system will be much simpler and can, therefore, be installed much more quickly."[28]

Much of the rest of Murphy's testimony described the political, economic, and technical advantages of geosynchronous satellites. He requested that the nation not commit itself to using medium-altitude satellites before examining the advantages of geosynchronous satellites. Murphy's competitive, perhaps even combative, attitude apparently struck a chord in the assembled members of Congress. They expressed their appreciation of his confidence but pointed out that General Bernard A. Schriever of the Air Force and Dr. Dryden of NASA had both expressed their opinions that geosynchronous satellites were some distance in the future. Murphy tried to explain that compared with Advent, Syncom—even the Syncom *Mark II*—was extremely simple. He also tried to explain that Syncom already used a traveling-wave tube amplifier and not the obsolete triode amplifier of Advent. Other objections raised were the limited station-keeping ability of compressed gas thrusters and the BPO's concern that messages from London to Australia would require two hops, with the resulting problems of delay and echo. The ending comments were almost prescient but were probably just observations. One was the suggestion that this system would be equally useful for coast-to-coast communication in the United States. Another was the suggestion that this system would be useful to the military. Domestic communications satellites (Domsats) would be delayed another decade, but the commercial-military trade-off would be made again in the following year.[29]

On September 18, 1962, Jaffe presented the NASA communications satellite program to the subcommittee. This program had been finalized in late 1961 and early 1962, before passage of the Communications Satellite Act and before the launch of *Telstar 1*. As described in the March NASA Communications Program Review, this program contained continued research efforts in passive satellites, medium-altitude active satellites, and geosynchronous active satellites. The committee members were not interested in passive satellites. They were interested in active satellites. When pressed as to whether Relay was better than Telstar or was simply a duplication, Jaffe tried to explain that Relay was superior only in (RF) power but that the performance of *Telstar 1*, especially its reliability, was a pleasant surprise. Jaffe had to explain that Comsat would choose its own system but that NASA would assist—especially through its R&D program. When pressed, Jaffe said that if the decision had to be made then, he would have to choose either Telstar or Echo—the only proven systems. The session ended with Jaffe assuring the subcommittee that NASA had sufficient funding to complete the current program of R&D.[30]

When opening the September 21, 1962, session, Committee Chairman Hechler commented on the probability that Syncom would provide another first for the U.S. space program. He pointed out that NASA, in the person of Jaffe, had considered it unlikely that any operational system could be attained in two years. Why, then, was DoD rejecting GEO and committing itself to an interim medium-altitude system? Eugene Fubini, deputy director of defense research and engineering, pointed out in his testimony that the military services needed communications in the polar regions. These needs could not be served by GEO satellites. Fubini was generally pessimistic about all space systems. He felt that reliability and multiple launches were the key to success in satellite communications.[31]

On September 27, 1962, representatives of the Department of State and USIA made presentations. They both emphasized the prestige effect that *Telstar 1* had generated. Both attributed much of this prestige to the television demonstrations. These were seen by many people in Western Europe. USIA pointed out that more citizens of the United Kingdom had heard of *Telstar 1* in 1962 than had heard of *Sputnik 1* in 1957. The USIA representative suggested that the cooperation of the French and British in the Telstar program had increased interest among the citizens of those countries.

In his concluding remarks, Dickieson, the last speaker on the last day, October 4, 1962, pointed out that there might very well be more than one system in the future. This was another prescient remark, but one not destined to be fulfilled for some time. The subcommittee concluded its hearings, impressed by the Hughes presentation on GEO but also conscious of the immediate success of *Telstar 1*.[32]

On October 4 President John F. Kennedy named the thirteen temporary directors of the new Comsat Corporation: Edgar F. Kaiser, president of Kaiser Industries; David M. Kennedy, head of Continental Illinois National Bank and Trust; Philip L. Graham, publisher of the *Washington Post;* Sidney J. Weinberg, partner in Goldman, Sachs; Bruce G. Sundlun, a Washington, D.C., lawyer; Byrne L. Litschgi, a Florida lawyer; Beardsley Graham, an aerospace engineer; Leonard Woodcock, vice-president of the United Automobile Workers; Sam Harris, a New York lawyer; George J. Feldman, another New York lawyer; Leonard H. Marks, a Washington, D.C., lawyer with Federal Communications Commission (FCC) experience; John T. Connor, president of Merck; and George L. Killion, president of American President Lines. The board consisted of five lawyers, three company presidents, two financiers, one publisher, one labor representative, and one engineer. Their first task was to incorporate; their second job was to issue stock—perhaps as much as $500 million worth.[33]

On October 22 the thirteen incorporators met for the first time. After confirming Graham as chairman, they were briefed by representatives of NASA, the Justice Department, the White House, the Bureau of the Budget, and the FCC.[34] At their second meeting on November 15 they were briefed by companies working on communications satellite experiments.[35] Both sessions were closed. The *Washington Post* columnist Drew Pearson, in two columns, gently chided the incorporators for some of their behavior and outlined their plans and problems.[36] The incorporators needed to draw up bylaws and proceed with incorporation; they needed to get financing; they needed to hire staff; they needed to deal with foreign "partners." Some of the flavor of their discussions can be inferred from a *Newport News (Va.) Times-Herald* article, which estimated the corporation's capital needs at $1 billion, the cost of forty medium-orbit satellites, including launch vehicle, at $200 million, and the cost of twenty ground stations at $240 million.[37] In addition, bankers were expecting fees of 7.5 percent to 10 percent for handling the stock issue.

Relay 1 Launch: December 1962

At 7:30 P.M. EDT on December 13, 1962, nineteen months after RCA won the Relay competition, *Relay 1* was launched by a Thor-Delta into a 712-by-4,020-nautical-mile orbit inclined 47.5 degrees to the equator. RCA had launched two TIROS (Television and Infra-Red Observation Satellite) satellites before the Relay award and another four before the *Relay 1* launch. All except *TIROS 1* (Thor-Able) were launched on the Thor-Delta. RCA was thus an experienced builder of Delta-class satellites. This experience was evident in the lightweight structure, the attitude-control system, and the use of redundancy to increase reliability. *Relay 1* also had almost five times the RF power of *Telstar 1*. This allowed the use of much smaller ground-station antennas. The Telstar TWTs were sealed into a foam-filled inner canister. The Relay TWTs were mounted on foam but were not sealed into a canister. Neither *Telstar 1* nor *Relay 1* had sufficient solar array power to operate the transponder. Transponder operation required battery power and was thus limited to a small fraction of the orbit. A transponder fault on *Relay 1* resulted in a low-voltage condition after launch. The transponder was on and could not be commanded off. After many days, a "work-around" was developed and communications experiments began. In addition, it was quickly discovered that the *Relay 1* n-on-p solar cells were dramatically inferior to the *Telstar 1* p-on-n cells. *Relay 1* had only half power after one year while *Telstar 1* had two-thirds power after two years. This was corrected in *Relay 2*.[38]

The *Relay 1* test results were similar to the *Telstar 1* results. Communications testing included standard voice and video tests. The biggest difference

Relay satellite. (Courtesy NASA)

was that additional Earth stations—in Brazil, Italy, Japan, and Germany—joined in the testing. Many of these had "System Gain/Noise Temperature" (G/T) characteristics that were not compatible with *Telstar 1* but were adequate for the 7dB-more-powerful *Relay 1*. The range of G/T was considerable, from 50 dB/K for the Andover and Pleumeur Bodou stations to 22–28 dB/K for the ITT-built stations. Only the MASER-receiver stations—Andover and Pleumeur

Bodou—and the Goonhilly station, with a G/T of 38 dB/K, were able to satisfactorily test television and 300-channel telephony. The environmental radiation test confirmed the *Telstar 1* results. In spite of the initial transponder problems, *Relay 1* had provided further proof that medium-altitude communications satellites were an available, reliable technology.

Syncom 1 Launch: February 1963

On January 29, 1963, NASA held a Syncom press briefing. The salient point, emphasized by Jaffe, was that this was primarily an experiment to determine the ability of NASA and its contractors to put a satellite into GEO. In addition to the three Delta stages, a fourth stage—the apogee kick motor (AKM)—would be controlled by the satellite itself. Syncom was a joint NASA-DoD program. The space segment was provided by NASA, but the ground segment was provided by DoD: two Army communications stations, at Camp Parks, California, and Fort Dix, New Jersey; and a shipboard facility provided by the Navy. USNS *Kingsport* would be in the harbor at Lagos, Nigeria, and would be critical to the firing of the AKM.[39]

Syncom 1 was launched into a nominal 150-by-19,312-nautical-mile GTO at 12:35 A.M. EDT on February 14, 1963. For several hours the satellite was tracked and various subsystems were tested. All was well. About five hours after launch, the AKM was fired. There was no communication with the spacecraft after that time. During the next few days, optical observatories searched for the spacecraft. On February 24, 1963, the Harvard College Observatory Boyden Station, located in Bloemfontain, South Africa, photographed the satellite. Additional photographs taken on February 25 and March 1 allowed an orbit to be calculated. *Syncom 1* was determined to be in a slightly subsynchronous orbit. The AKM had worked, but something on the satellite had failed.[40]

The first two years of the Kennedy administration had seen a complete change in the political environment surrounding communications satellites. At the end of the Eisenhower administration, there had been a widely held assumption that industry, specifically AT&T and possibly others, would be responsible for the development and implementation of satellite communication systems. There had been some reluctance on the part of NASA executives to allow industry to develop communications satellites and further reluctance by political appointees to be seen as supporting AT&T, but all seemed resigned to a commercial international communications satellite system developed, implemented, and dominated by AT&T. Within a few months after Kennedy's inauguration, AT&T was told that it would not be allowed to "preempt" the government's program in space. A month later the president had committed the United States to developing a global communications satellite system. The White House and Congress then proceeded to pass a law that granted a new government-formed entity the monopoly right to build and operate the *single* global communications satellite system.

Before the Kennedy inauguration, the military had dominated the field of satellite communications. The SCORE experiment launch, the Courier satellite launch, and the ongoing Advent program seemed far ahead of AT&T's Echo experiment and the breadboard subsystems being tested by AT&T and Hughes for satellites one-tenth the size of Advent. Within two years, two completely different medium-altitude designs, *Telstar 1* and *Relay 1,* had been successfully launched and operated. Development of the Hughes Syncom satellite was being supported by NASA, although the first launch had not been successful. DoD commitment to the geosynchronous Advent program had shed more-favorable light on GEO, until the cancellation of the Advent program in the face of huge overruns and the obvious success of the two MEO satellites. NASA and MEO had overtaken DoD and GEO just as the government had overtaken industry.

The successful *Telstar 1* launch on July 10, 1962, proved that medium-altitude communications satellites were eminently practical. This did nothing to improve AT&T's chances of operating such a system; by this time, the congressional bill clearly would not allow AT&T to do so. Within a few weeks the question became: What kind of system should the new organization (Comsat) launch? Jaffe, communications director at NASA Headquarters, was quoted as saying that a Telstar-type satellite (i.e., medium altitude) would be the first system because Syncom had not yet proved its reliability.[41] This comment was made less than six months after Jaffe's boss, Stoller, had claimed that GEO was best. If the experts were undecided, what chance was there that the lawyers in Congress and the lawyers on the board of the new organization would be able to choose wisely?

6. Choosing a System

NASA favors 22,300-mile-high [GEO] satellites. —***Missiles and Rockets,*** **April 2, 1962**

NASA sees Telstar-type [MEO] satellite as best. —***Aviation Week and Space Technology,*** **September 24, 1962**

Of the thirteen incorporators named by President John F. Kennedy in response to the provisions of the Communications Satellite Act of 1962, only one had any significant technical background. Beardsley Graham, an aerospace engineer, had worked at Stanford Research Institute and at Lockheed. While at Lockheed, he had been active in satellite communications studies—done by Lockheed itself and in partnership with GTE and RCA. In an interview published in January 1963, Graham outlined the issues before the incorporators and the board of directors who would follow them.[1] He stated that the conclusion of the Lockheed study was that the main issues were not technical but were rather business, regulatory, and international issues. Graham was convinced that geosynchronous systems would be in place in the fairly near future but that, regardless of the system chosen, the new corporation would have a lot to discuss when formal meetings began. In response to the interviewer's assumption that the incorporators were simply supposed to locate a board of directors and resign, Graham pointed out that the incorporators were charged by law to act as a board of directors until their successors were elected. Before such an election, stock would have to be sold. Before stock could be sold, the corporation would have to describe its new business in sufficient detail to satisfy investors and the Securities and Exchange Commission (SEC).

The new communications satellite entity was incorporated in the District of Columbia on February 1, 1963, as the Communications Satellite (Comsat) Corporation. Before that date, Philip Graham, the chairman appointed by President Kennedy, had resigned.[2] On February 28, 1963, Comsat announced that Leo D. Welch of Standard Oil Company (New Jersey) had been named chairman and that Joseph V. Charyk, undersecretary of the Air Force, had been named president of the organization. Welch's background was in international finance; Charyk's background was highly technical and included significant experience in the reconnaissance satellite business. The salaries for the two men were to be $125,000 for Welch and $80,000 for Charyk per year. This was a significant pay raise for Charyk, who earned $20,000 yearly as undersecretary of the Air Force. The Federal Communications Commission (FCC) authorized the corporation to borrow up to $5 million from banks. The banks included Continental Illinois National Bank, of which David Kennedy, an incorporator, was an officer.[3]

In addition to the standard problems of getting an organization running, Comsat had three major concerns: (1) keeping Congress happy, (2) keeping the Europeans happy, and (3) trying to determine what the eventual operational system would look like. Congress was unhappy with the high salaries paid to Welch and Charyk, their luxurious offices in Tregaron, a Washington mansion, and the general uncertainty involved in the enterprise.[4] The Conference of European Postal and Telecommunications Administrations (CEPT) formed a committee in December 1962 to study the issue of joining a U.S.-led global communications system. Though recognizing that the majority of international telecommunications traffic originated or terminated in the United States, the Europeans were eager to gain maximum control and make equipment sales.[5] *Telstar 1* and *Relay 1* had been launched relatively successfully, but *Syncom 1* had failed after injection into GEO (geosynchronous Earth orbit). With the major exception of analysts at AT&T, most felt that the geosynchronous satellite held the greatest promise. The RAND Corporation had issued a study of system choices in February 1963.[6] The principal author was Siegfried Reiger, who became a Comsat employee within a few months. Joining Comsat at the same time as Reiger were Sidney Metzger of RCA and Edwin J. Istvans of the Air Force. By the end of the year, Comsat had personnel from the Air Force, AT&T, NASA, and RCA but not from Hughes Aircraft Company.

NASA

The June 29, 1963, NASA Applications Program Review was NASA Administrator James Webb's first formal meeting with Robert Garbarini, the new director of the Office of Applications.[7] Milton Stoller, the previous director, had recently died, at a very young age. The 1962 Program Review had been dedicated to communications. At this review, communications was presented as just one of the three components of the Applications Program, which also included weather and future applications.

Leonard Jaffe made the communications systems presentation. He emphasized the changes that had taken place since the last program review, in March 1962. According to Jaffe, "The intermediate altitude [MEO] active satellite appeared then, as it does now, to offer the most promise for an early operational system."[8] In 1962 NASA had intended to continue its MEO (medium Earth orbit) R&D with an advanced system that would be the forerunner of an operational system. Since that time, the Department of Defense (DoD) had decided to launch its own MEO system, and Comsat was also apparently preparing to launch an MEO system. NASA had therefore abandoned its MEO program and decided to emphasize GEO satellites. The Echo 2 program had been kept, but the Rebound passive program had been canceled, as had also the S-64 geosynchronous radiation experiment. *Telstar 2* had recently been launched (on May 7, 1963), and *Relay 2* would be launched at the end of the year. The rest of the NASA communications systems program would be geosynchronous. *Syncom 2* would be launched in the following month (on July 26, 1963) and *Syncom 3* in the fourth quarter. The new programs were Advanced Syncom, scheduled for launches in 1964, and Advanced Systems, scheduled for launches in 1967.[9]

The Telstar and Relay satellites had problems but generally performed well. Jaffe pointed out that *Telstar 1* had lasted only seven months, whereas *Relay 1* was still healthy and about to exceed the *Telstar 1* lifetime. Furthermore, the higher power of the Relay transponder allowed smaller antennas to participate in the experiments. There were those who criticized NASA for spending taxpayer funds when industry was willing to spend its own money; NASA countered this complaint by pointing to the superiority of its chosen system, Relay, over the privately funded, AT&T-built Telstar. *Telstar 1* had succumbed to radiation damage to its transistors after 186 days. At least some of the radiation damage was due to the *Starfish* nuclear explosion, which had greatly increased radiation levels in the van Allen belts just before the launch of *Telstar 1*. A power-controller fault linked to high temperatures had initially prevented the operation of the Relay transponder. Eventually a means of "working around" the problem was developed, and *Relay 1* had been operating successfully for 178 days.[10] Several spurious commands had registered in the command system. It was not clear what had caused these. In addition one of the battery strings had suffered failures. Finally, Relay's n-on-p solar cells were inferior to Telstar's p-on-n cells and would be replaced on *Relay 2*. *Telstar 2* had been in orbit for 45 days and appeared to be healthy. Jaffe listed eight major results of the MEO program:

1. Space Communications *technology* is adequate.
2. Space Communications has been *demonstrated* in MEO.
3. Space Communications *ground stations* now exist.
4. Ground station technology is *international.*
5. Accurate *antenna pointing* has been demonstrated.
6. *Orbit information* has been routinely derived and disseminated.
7. Space Communications can *share frequencies* with terrestrial microwave.
8. *Coordination* of ground stations has been demonstrated.[11]

The most important question that had not yet been answered by the MEO programs was whether satellite lifetimes were sufficient to permit economically profitable use.

Syncom 1 had failed after launch, but even it had produced some successes. The communications payload had been turned on before the AKM firing and had been working. Optical sightings showed that the satellite had been injected into synchronous orbit. Analysis and testing had shown that the probable culprit was the nitrogen tanks, which were made from a brittle titanium alloy. Several changes would be made in *Syncom 2* to correct this problem.[12]

Advanced Syncom was currently in the study stage. The study contract had been due to end April 30, 1963, but had been extended to August 31, 1963, to allow NASA to observe the performance of the basic Syncom before committing to an Advanced Syncom construction program. This program would have a high-gain electronically despun antenna, several linearized multiple-access transponders, a hot-gas propulsion system, and many other improvements.[13]

Ground stations were being built for future use by several countries. DoD, in Project Trade Post, had begun using transportable ground stations to operate with Telstar. NASA was installing a second 85-foot antenna at Rosman, North Carolina, to use with future NASA communications programs. NASA thought that commercial antennas might be dedicated to operational systems in the near future, leading to a requirement for NASA to have its own ground stations. NASA was also considering upgrading the Mojave, California (Goldstone), station to support television transmission of the 1964 Olympics in Tokyo.[14]

Jaffe also outlined the Comsat plans to which NASA was privy. Comsat's first goal was to contract for several system studies in 1963. The results of these would help it to decide on the design of an operational global communications satellite system. Comsat hoped to let contracts for this system in July 1964 and to have the system operational in 1967. In November the International Telecommunications Union was expected to designate the terrestrial common-carrier frequencies (C-band: 6 GHz up and 4 GHz down) as shared frequencies with communications satellites.[15]

Hugh Dryden and Robert Seamans were both somewhat cautious regarding Jaffe's presentation. Seamans pointed out that the success of *Syncom 2* was not sufficient to guarantee implementation of Advanced Syncom, especially given the congressional view that Comsat should pay for its own R&D. Seamans also pointed out that interest in gravity-gradient stabilization techniques might influence the Advanced Syncom program. Webb tried to cheer up Jaffe and the staff of the NASA Goddard Space Flight Center by stating that it was preferable for him to be in the position of restraining their enthusiasm rather than trying to generate enthusiasm. In this exchange and in a later discussion about the importance of transmitting the Olympic games from Tokyo, a strong current of disagreement ran between Dryden and Harry Goett, the Goddard director. Dryden was much more conservative. Goett would eventually leave Goddard and NASA.[16]

In the discussion on funding, there were several questions about possible reimbursement by Comsat. Congress was not interested in funding Comsat R&D. Seamans pointed out that Jaffe's Advanced Communications System, a high-powered system for mobile communications, had gone over to the U.S. Bureau of the Budget as the Advanced Technology Satellite and had included the synchronous meteorological satellite and gravity-gradient experiments.[17]

Comsat

NASA was invited to participate in internal Comsat discussions—and accepted. Jaffe, head of the Communications Office, and Lieutenant Colonel Robert E. Warren attended.[18] Comsat acquired the services of two consultants in July: Milton U. Clausner, a member of the U.S. Air Force Science Advisory Board (SAB), and J. P. Ruina, a former director of the Advanced Research Projects Agency. At the same time it prepared to issue study contracts to ITT, GTE, RCA, and AT&T. There was no lack of analytical capability applied to the problem of choosing a system for Comsat. Adding to the options available to Comsat was GEO—finally reached by *Syncom 2* after launch in July. *Syncom 2* had proven that GEO was attainable, but the light weight of the satellite and the much longer communications path length constrained the communications payload to a single telephone circuit—significantly less than the wideband capability of Relay and Telstar.

The FCC in July made public its concern that Comsat no longer had "definite plans for an early issue of stock." The appointed Comsat board was in the position of making decisions about the future of the company—decisions that should have been made by the shareholders. Comsat responded to the criticism by suggesting that decisions had to be made about the system configuration before issuing stock. Until a basic program was outlined, it was unclear how much equity capital should be acquired, and it was unclear how to categorize risks—a legal requirement.[19]

In September 1963, articles by Charyk and Reiger appeared in the AIAA's (American Institute of Aeronautics and Astronautics) *Astronautics and Aerospace Engineering.*[20] The articles outlined the proposed international cooperation—locally owned Earth stations and a jointly owned space segment—and the technical problems and choices confronting Comsat. The technical choices involved spacecraft orbit and attitude control. Reiger summarized the orbit selection problem by stating: "Most engineers agree that a random system with satellites of the general type of *Telstar* and *Relay*, could be placed in operation at an earlier date than a [geo]stationary system. . . . But there also appears to be general agreement that in the long run the [geo]stationary-satellite concept offers the greatest promise and growth potential." Attitude control could be uncontrolled, spin-stabilized, or fully stabilized (three-axis stabilized). Both Reiger and Charyk pointed to gravity-gradient stabilization as an interesting and potentially fruitful technique (it was a military favorite). The technical problems were in the communications area and included the time-delay and echo-suppression problems associated with geostationary satellites and the problems that resulted when several carriers (e.g., from several Earth stations) were simultaneously amplified by the satellite's traveling-wave tubes (TWTs). Reiger stated that there was insufficient geostationary satellite experience to evaluate the problems associated with this orbit.

Within a year of receiving a sole-source contract from NASA to build three Syncom satellites, Hughes Aircraft Company engineers had begun studies of an Advanced Syncom. The first three Syncom satellites would prove the practicality of the spacecraft bus, but the communications payload was not very impressive. Advanced Syncom, launched by the proven Atlas-Agena, would weigh about 750 pounds in GEO. The payload would support thousands of voice circuits rather than the "single-voice circuit" of the earlier *Syncom 1* and *Syncom 2*. The TWTs would be higher-power, and the antenna would direct most of its energy onto Earth's surface. Finally, the satellite transponders would have a multiple-access capability: more than one station could use the same transponder at the same time. A description of Advanced Syncom was in the same issue of *Astronautics and Aerospace Engineering* as the articles by Comsat's Charyk and Reiger.[21] Advanced Syncom had also been mentioned earlier at a "Technical Background Briefing" on the Syncom project in January 1963 before the launch of *Syncom 1*.[22]

Congressional objections to the spending of taxpayer dollars for the benefit of a private corporation, Comsat, made NASA's job particularly difficult. In an effort to allow more time for *Syncom 2* to prove itself and to convince Congress of the benefits of Advanced Syncom, NASA extended the study contract for another two months in June.[23] At about the same time, Hughes proposed an Intermediate Syncom.[24] This satellite would be lighter and less costly. It would also be useful to DoD and the Apollo program.

The successful July 26, 1963, launch of *Syncom 2* made it clear that geosynchronous satellites were a reality. The complex station-keeping problem was being solved—apparently easily. On July 27 the hydrogen-peroxide jets (hot-gas thrusters) were successfully used to reverse the 7 degrees/day eastward drift of the satellite. On July 31 thrusters were successfully used to precess the satellite from its in-plane attitude to an orbit-normal attitude, which would optimize communications performance. On August 11 the westward drift was slowed using pulsed hydrogen-peroxide thrusters again, and on August 12 the nitrogen (cold-gas) thrusters were used to reduce the drift further. A series of further maneuvers reduced the drift to less than 0.1 degree/day. The satellite was positioned over the Atlantic, ready to carry communications traffic.[25]

Hughes was extremely active in attempting to find additional markets for Syncom. One idea that was resurrected—it had been D. D. Williams's original suggestion in 1959, before Harold Rosen convinced him that communications was a better application—was to use Syncom as a navigational system.[26] More immediate was the attempt to convince DoD that the Hughes Synchronous Altitude Communications Satellite System (SACS) had better performance than a Medium Altitude Communications Satellite System (MACS). On September 19, 1963, "Pat" Hyland, the Hughes general manager, met with DoD's Fubini but was unsuccessful in his attempts to gain a convert.[27] More productive were Hyland's discussions with Bob Gilruth, of NASA, to whom he suggested that Syncom be used to provide reliable communications links between the widespread tracking stations of the Apollo project.[28] Attempts to find additional markets were only accelerated by the news that NASA would not pursue the Advanced Syncom flight program. At a September 6, 1963, meeting, Webb told Garbarini, the NASA applications director, that he did not think Congress would allow NASA to fund hardware development because of the perception that NASA was subsidizing Comsat.[29]

After canceling Advent in the summer of 1962, DoD had decided to develop a smaller, lower-orbit satellite (MACS) for its Initial Defense Satellite Communications Program (IDCSP). In April 1963 Lieutenant General A. G. Starbird, director of the Defense Communications Agency (DCA), described the program to Congress, and in May study proposals were solicited from industry. Two study teams were funded: one led by General Electric, the Advent contractor, teamed with Motorola; and another led by Philco, the Courier contractor, teamed with STL/TRW.[30]

Comsat objected to MACS on the grounds that the Communications Satellite Act of 1962 had given it total responsibility for U.S. satellite communications. On October 11, 1963, Secretary of Defense Robert S. McNamara wrote to Comsat's president, Charyk, suggesting that Comsat might provide both commercial communications services for DoD and unique communications services. Many in DoD assumed that Comsat was too busy to get involved in a separate DoD system. Charyk responded on October 26, 1963, that Comsat was interested but that more time was needed to study the issues. DoD was forced to negotiate with Comsat. The military expressed a preference for the proven random, medium-orbit system, preferably using gravity-gradient-stabilized satellites. To preserve DoD's options, on November 15, 1963, Philco's study contract was extended by DoD. General Electric's contract was not extended. Philco realized that it had won the competition but still might not get a contract.[31]

Comsat's argument was that DoD would receive far more service at a lower cost by joining with Comsat. DoD countered with arguments about foreign participation and special needs. Wilbur L. Pritchard, Aerospace Corporation group director for communications satellite systems, told a House subcommittee that Comsat would end up charging DoD as much as DoD would have to pay to put up its own system. The Army seemed more willing to support GEO systems than did the Air Force or the Navy, but the Army doubted that DoD would surrender the flexibility to optimize the system for its own needs—flexibility that leasing from a civilian organization would entail. In addition, the military preferred the two 50-MHz allotments in the 7/8-GHz band (Syncom used 8/2 GHz) rather than the two 500-MHz allotments available in the "civilian" 6/4-GHz band selected by Comsat. As a result of these discussions, Senators Albert Gore (D, Tenn.) and Ralph Yarborough (D, Tex.) called for the repeal of the Communications Satellite Act of 1962 on the grounds that Comsat's primary customer would be the U.S. government.[32]

Comsat made several announcements at the end of 1963. One of these made public a letter from James E. Dingman, executive vice-president of AT&T, to Leo Welch, chairman and CEO of Comsat. The letter contained AT&T's assessment that both cables and satellites would be necessary in the future, especially for diversity. The letter seems to suggest that AT&T preferred cables and would use them wherever it could but that, in many cases, providing *initial* service via satellite would be easier and that, in all cases, providing satellite capability as a backup to cable service would be prudent.[33] The second announcement outlined Comsat's near-term plans. Comsat issued a Request for Proposal (RFP) for a satellite design, either MEO or GEO, which would constitute Comsat's "basic system." No decision on the "basic system" would be made until after the design(s) had been evaluated. Comsat also announced the possibility that an "early capacity" might be established consisting of "a synchronous satellite orbited on an experimental-operational basis in 1965," with a bandwidth and power that could "provide a capability for television or, alternatively, for facsimile, data, or telegraphic message traffic or for up to 240 2-way telephone channels."[34]

On February 25, 1963, NASA had announced its intention to concentrate on stationary satellites due to the recent incorporation of Comsat and the summer 1962 cancellation of DoD's Advent. DoD had signaled its intention to temporarily abandon GEO and put up an initial system consisting of medium-orbit satellites, probably with gravity-gradient stabilization. All of these decisions complicated Comsat's task—it was assumed that the initial commercial system would be medium-orbit. In December 1963, Hughes proposed a commercial version of Syncom; this version could be launched in early 1965 and serve both experimental and operational needs. The satellite would utilize the 6/4-GHz "commercial" frequencies as opposed to the 8/2-GHz "military" frequencies. In addition, the satellite would have a squinted beam that would produce a toroidal (doughnut) pattern aimed at the North Atlantic. For the "basic" system to be launched later, Hughes proposed the Intermediate Syncom (a smaller version of Advanced Syncom), which could be launched on the Delta rather than the Atlas-Agena. Without committing to GEO for the "basic" system, Comsat decided to accept the Hughes proposal and launch an "early bird" before making a final decision on system type.

The choice of communications satellite system was not easy. AT&T and DoD were convinced that medium-altitude satellites were best; many others were convinced that high-altitude geosynchronous satellites were best. In April 1962 Stoller, director of the NASA Office of Applications, had stated, "When all the technology is in hand, it will be the synchronous-orbit satellites which will be the most attractive to us."[35] On the other hand, in September 1962 Jaffe, director of the NASA communications group under Stoller, had suggested that a medium-earth-orbit system similar to Telstar would be best.[36] Reiger, the Comsat systems analysis manager, believed, as did many others, that a GEO system would be best but that there was simply insufficient experience with this type to ensure success.

Comsat convened a meeting on January 23, 1964 with Asher Ende, chief of the FCC's Office of Satellite Communications, and several members of his staff, as well as John J. Kelleher, from NASA's Office of Communications. Comsat saw *Early Bird* as both experimental and commercial. After a month or two of testing, NASA proposed to make the satellite available for commercial use. All agreed that FCC approval was required before construction could begin. The FCC was not comfortable with Comsat's intention to use the AT&T Andover station, preferring that Comsat own and operate its own Earth station.[37]

The first annual report of Comsat to the president and Congress provides an analysis of the Comsat program. The report included contracts awarded to AT&T, RCA, and Hughes to study the problems of multiple access—several Earth stations using the same transponder simultaneously. An RFP had been released on December 22, 1963, for commercial satellite design development. These proposals were due February 10, 1964. In addition, Comsat reported that it would launch one or more geosynchronous satellites in 1965 on an experimental/operational basis. The report also listed senior members of the staff. Interestingly, four were from RCA, three from the Air Force, two each from NASA and RAND, and one each from the Navy and AT&T.[38]

Four proposals were received in response to Comsat's RFP: AT&T teamed with RCA, TRW teamed with ITT, Hughes, and Philco.[39] The AT&T-RCA and the Philco proposals were for medium-orbit (6,000-mile-altitude) random systems. The TRW (STL)-ITT proposal was for a phased/controlled medium-altitude system. Hughes, as expected, proposed a geosynchronous system. The medium-altitude systems would be the most expensive. A $200 million stock offering was planned in the near future to cover the costs of implementing any of the systems. These satellites were to be launched in 1966, although it was becoming clear that a 1967 date might be more realistic. At the same time that Comsat was analyzing the technical aspects of the global communications satellite system, it was also dealing with the European nations and their interest in participating as equal partners in this enterprise.[40]

On March 4, 1964, Comsat requested permission from the FCC to launch *Early Bird* in early 1965 into geostationary orbit over the Atlantic. The launch would be on a Delta launch vehicle from Cape Kennedy. The 85-pound satellite would provide 240 voice circuits or one television channel. Part of the rationale for *Early Bird* was the success of *Syncom 2* (launched on July 26, 1963) and the resulting desire to experiment with a similar satellite using commercial frequencies (6/4 GHz) in a commercial environment. AT&T planned to lease 100 circuits on *Early Bird* to handle peak loads and diversity (cable replacement). On March 17, 1964, Comsat negotiated a $4–$8 million incentive contract with Hughes for two satellites and long-lead items for a third satellite. Comsat made it clear that this was an experimental proto-operational system, not the "basic" system that it would launch in 1967. In April the FCC approved the launch of *Early Bird*.[41]

The first NASA Program Review of the communications satellite program in 1962 had been a dedicated review. By the time of the second review in 1963, communications satellites had been combined with other "application" satellites. By the time of the third review in 1964, NASA management was too busy to listen for more than a few minutes to a discussion of communications satellites.

On June 25, 1964, Jaffe made his short presentation on Advanced Syncom to NASA management. He explained that the Advanced Syncom study program (1962–63) had preceded the Advanced Technology Satellite (ATS) study (1963–64) and was intended to study a follow-on to the original Syncom program. The most important part of this study was development of a phased-array antenna, which would provide much higher gain by concentrating the transmitter output on the surface of Earth. An engineering model of Advanced Syncom had been completed in February 1963. According to Jaffe, it was then realized that the spacecraft could be used for more than one mission. Jaffe didn't mention here that Congress was less than enthusiastic about supporting research that would benefit only Comsat. Nor did Jaffe mention the ATS meteorological payload—although it was included on his slides. The ATS program (later named the Applications Technology Satellite) had three phases. The first was a gravity-gradient-stabilized satellite in a 6,000-mile (10,000-kilometer) orbit. This would be followed by a spin-stabilized satellite in GEO. Last would be a gravity-gradient-stabilized satellite in GEO. The major payload experiments would be communications and a weather camera. In contrast to the ad hoc nature of the earlier programs, ATS would be carefully planned to examine all relevant technologies.[42]

Meanwhile, the commercial world was giving Comsat a vote of confidence. The first five million $20 shares of Comsat stock had been sold, exclusively to communications common carriers. Although the market handled the second five million shares, the FCC was responsible for apportioning the first sales. Over 200 carriers had notified the FCC of their interest by the March 23, 1964, deadline. By the official May 26, 1964, deadline for bids, AT&T had offered to buy $85 million of the $100 million worth of shares reserved for communications carriers. Because of the large oversubscription, AT&T was able to buy only $57.9 million, or 29 percent of the total shares. ITT bought $21 million worth (11 percent), GTE bought $7 million worth (4 percent), and RCA bought $5 million worth (3 percent). Other carriers bought the remaining 3 percent available for communications common carriers. Six of the directors would be elected by the communications carriers. Each holder of 8 percent of the total shares would earn a director's seat. AT&T thus had at least three seats guaranteed, and ITT probably had two seats. The public shares were offered on June 2, 1964, and were immediately snapped up.[43]

The June 2, 1964, prospectus provides a snapshot of Comsat at that time and a view of its plans for the near future. On June 1, 1964, Comsat had 79 employees and owed banks almost $2 million. Comsat planned to expend the $200 million it expected to receive from the stock offering on (1) an experimental-operational satellite(s), costing $14 million (low estimate), (2) designs and prototypes to establish the "basic" system, costing $55 million, (3) the "basic" system itself, costing $75 million, (4) Earth stations, costing $17 million, and (5) additional R&D and administrative activities. Comsat was also negotiating with the National Communications System (NCS) to provide military communications. Comsat applied to the FCC at this time for authority to own four U.S. Earth stations in the global system.

International arrangements had begun as early as 1960, when AT&T had first suggested to its "correspondents" in the submarine cable programs (e.g., TAT) that a satellite program was a possibility. NASA had continued these discussions (as had AT&T) in 1961 and 1962. While the Communications Satellite Act of 1962 was going through Congress, the Europeans were considering their options. They decided that they could maximize their benefit from the satellite program by negotiating as a bloc. In December 1962 the CEPT began formal studies to establish a basis for discussions with the United States. In July 1963 the European Conference on Satellite Communications met in London. After preliminary meetings with the U.S. State Department and Comsat, a plenary meeting of the conference was held in Rome from November 26 to 29. The Europeans proposed that they set up a counterpart to Comsat. At later informal meetings a consensus was reached that a consortium was probably the best approach, rather than the series of bilateral agreements preferred by the State Department. A London meeting in April explored the details of the consortium approach and proposed that Comsat be the manager of the consortium. A London drafting meeting was held in May, and final meetings were held in Washington on July 21-25, 1964. The resulting agreement was opened for signature on August 19, 1964. The structure of the International Telecommunications Satellite Organization (Intelsat) had been determined essentially in the form in which it would exist for the next several decades.[44]

A series of meetings in July 1964 determined that a U.S. military communications system could not be owned and operated by an organization partly owned by foreign entities. At a press conference on July 15, 1964, Defense Secretary McNamara announced that DoD would build its own satellite communications system. The military portion of the NCS was not to be a Comsat function. Both Welch and Charyk attempted to keep the door open in letters to McNamara, but at least for the near term, Comsat was out of the MACS program and no longer partially committed to a MEO program.

Syncom 3

Hughes was beginning to pick up momentum in the competition to be the primary provider of communications satellites. In 1964 it began its contract for two GEO satellites (HS-303, *Early Bird*) for Comsat. In March NASA had awarded Hughes a contract for five Advanced Technological Satellites (ATS). And in August the Hughes *Syncom 3* was launched into geostationary orbit. Hughes still had not been awarded a patent for the Syncom orbit- and attitude-control techniques, but this was sure to come.[45]

On August 19, 1964, *Syncom 3* was launched on a Thrust-Augmented-Delta (TAD). This Delta had three "strap-on" solid rockets attached to the first stage. The additional thrust of the first stage allowed part of the second-stage thrust to be used to reduce the inclination of the geosynchronous transfer orbit from 28 degrees to 17 degrees. The remaining inclination was removed when the AKM was fired. Unlike the previous Syncom launches, the satellite had to be reoriented to ensure that the thrust vector from the AKM pointed in the proper direction—no longer simply pointing in the direction of flight. This was an amazing achievement and further evidence that the Hughes orbit- and attitude-control scheme was superior. The inventor of this scheme, D. D. Williams, was singled out by Rosen as one of the "major contributors" to the success of the program.[46]

Syncom 3 was used to transmit television from the 1964 Olympic Games in Tokyo. The U.S. Navy Earth station at Point Mugu, California, had been modified for the telecasts, and Japan had built/modified a station at Kashima, northeast of Tokyo, to transmit the military X-band frequencies used by Syncom. The transmission was not "commercial quality," but no one really noticed. NBC complained that the pool film provided by the Japanese broadcasting company NHK was insufficient and was sent late and that very little live coverage was provided. A more humorous, but perhaps nastier, complaint was part of a poem by Peter Dickinson in the September 1964 issue of *Punch:*

Syncom
The Pad is hidden in chemical smokes and flames.
    The rocket shudders, hesitates, starts to rise,
    Then soars—and seems in its soaring to symbolize
And justify man's most extravagant claims.
There goes the pride and power and resource that tames
    The black implacable spaces beyond our skies.
    Lords of the Universe! Now we can televise
The 1964 Olympic Games.

Well, and why not? What hope was there of applying
    Those mountains of cash and cleverness to aid,
For instance, nations oppressed or children dying?
    Wouldn't man's pride, etcetera, have made
Quite certain that these resources were spent on a ploy
As beautiful, novel and useless as this huge toy?

Syncom 3 satellite in foreground; U.S. Navy antenna at Point Mugu, California, in background (*note:* size of the satellite is greatly exaggerated by the perspective effect). (Courtesy Hughes/BSS)

Many wondered at the time and later if space was the right place to spend mankind's resources. Some, like the writer of the *Punch* poem, could not discriminate between the practical and the prestigious.[47]

It is easy to forget that Syncom was a "dual-use" program. The program was jointly sponsored by NASA and DoD. After the NASA experimentation had been completed, DoD was interested in using the spacecraft to provide communications services over the Pacific. The year of solar minimum came to an end with no military satellite having been launched to supplement HF radio. NASA was ready at the end of 1964 to begin transferring the spacecraft and associated TT&C facilities to DoD. The ground stations had always been provided by DoD. By mid-1965 the transfer was complete. *Syncom 3* was providing services over the Pacific, and *Syncom 2* was providing service over the Indian Ocean.[48]

Comsat: New Board Members

The Comsat stock offering in June was very popular—so popular that the size of purchases was limited. One result was that Comsat stock ownership was dispersed. The average public shareholder held only 27 shares. Of the 130,000 shareholders, about 120,000 held less than 100 shares. Two other results were the upward pressure on the stock price ($48 per share by mid-August) and the lack of a focus for challenges to management. At the first shareholder meeting on September 17, 1964, the management nominees were elected to the board with little difficulty: Leo Welch (chairman of Comsat), Joseph Charyk (president of Comsat), David M. Kennedy (head of Continental Illinois National Bank), George L. Killion (president of American President Lines), Bruce G. Sundlun (a lawyer), and Leonard H. Marks (another lawyer). President Lyndon Johnson nominated the three government directors: Clark Kerr (president of the University of California), George Meany (president of the AFL-CIO), and Frederick Donner (chairman of General Motors). The communications carriers directors were Harold M. Botkin, James E. Dingman, and Horace Molton of AT&T; Eugene R. Black and Ted B. Westfall of ITT; and Douglas S. Guild of Hawaiian Telephone. Comsat was now operating as a commercial venture with equity funding and a board elected by the shareholders.[49]

The congressional hearings on the military use of Comsat—held in March, April, May, and August 1964—left many members of Congress with a bad taste. In October the military operations committee reported that the effort to merge the military and the civilian systems was "ill-advised, poorly timed, and badly coordinated." The committee wanted DoD to search for economies but felt this was clearly not one of them. DoD was directed to proceed with the medium-altitude system that it had planned in 1962 and for which Philco was apparently the winning contractor. The committee members did agree that the Titan III launch vehicle might be useful, as would gravity-gradient experiments, but most of their recommendations related to improving management of the NCS and national telecommunications policy.[50]

On January 25, 1965, Comsat reopened the issue of providing a communications satellite system for the military. Twenty-four Hughes satellites, similar to *Early Bird,* were to be launched, eight at a time, on three Titan IIIC launch vehicles. Comsat had blamed the Pentagon for the previous debacle, but now it wanted to try again. Philco, the winner of the 1963 DoD competition, had not been able to proceed for over a year because of Comsat's blocking moves. Philco was annoyed that the process had started all over again. Comsat proposed to offer a sole-source contract to Hughes. Philco objected to the FCC that Comsat legislation required competitive bidding. On February 23 the FCC blocked Comsat's move: competitive bidding would be required. Eventually, Philco would build the IDCSP satellites, but politics had delayed launch from 1964 to 1966.[51]

Early Bird (Intelsat 1)

Comsat was bullish on Hughes, but it still wasn't clear that the Comsat "basic" system would be GEO. Syncom had been successful—and orbit and attitude control appeared to be much simpler than originally thought—but the telephone companies had (generally) only experimented with the MEO satellites Telstar and/or Relay because of the military frequencies used by Syncom. An article in the aerospace press speculated that the decision for GEO had been made—along with the decision to use the $3.5 million Delta launch vehicle—but Comsat did not confirm this. The stock, though still selling for almost three times the initial price, was constantly moving up and down because no one knew what was going to happen. *Early Bird* was thus eagerly anticipated. AT&T wanted to use 100 (of the 240) circuits, Canada wanted to use 24, and Britain, France, and Germany were eager to participate, even without multiple-access capability. The telephone companies were willing to take turns using the satellite.[52]

NASA was responsible for the Delta launch from Cape Kennedy, but except for basic tracking services, NASA was just a spectator after launch and had no responsibilities after the firing of the AKM. All orbit-determination data would be collected by the Andover station, which also transmitted all commands. Commanding would be at the 6-GHz communications frequencies rather than the NASA-standard VHF frequencies. All orbit-determination and -control functions would be performed by Comsat personnel. NASA provided the "launching and other services" prescribed in the Communications Satellite Act, but Comsat was in charge. As scheduled in late 1964 and early 1965, *Early Bird* would be placed in a zero-inclination GEO at 30 degrees west longitude on March 1,1965.[53]

Both NASA and Comsat provided press releases and briefings describing the upcoming launch and the key officials involved in the launch. The Comsat fact sheet listed Comsat, NASA, Hughes (the spacecraft manufacturer), AT&T

(owner of the Andover Earth station), and Douglas (the launch vehicle manufacturer) personnel—leading off with Siegfried Reiger, Comsat's vice-president for technical. The NASA mission operations report listed only NASA personnel—leading off with Leonard Jaffe, NASA's director of communications and navigation programs.[54]

*Early Bird (Intelsat 1)*was scheduled for launch on April 6, 1965, on a thrust-augmented Delta (TAD). This would be the thirtieth Delta launch. The spacecraft was almost identical to *Syncom 3* but used the commercial 6-GHz and 4-GHz frequencies rather than the 8-GHz and 2-GHz military frequencies used by the Syncom satellites. Like *Syncom 3*, *Early Bird* would be geostationary, not just geosynchronous. The satellite's orbit would be in the plane of Earth's equator, not inclined to it. Satellites that are fired east from Kennedy Space Center (latitude 28.5 degrees north) cross the equator at an angle of 28.5 degrees. This angle, the orbital inclination, will result in a geosynchronous satellite moving from north to south in a narrow figure-eight pattern. Inclination can be removed when Δv is added to the spacecraft at perigee, about twenty minutes after launch, or at apogee, when the AKM is fired. In either case, the pointing attitude of the satellite must be changed. The attitude change required for this "dogleg" maneuver is one of the reasons geostationary orbit was considered so difficult to achieve. It was planned that about 10 degrees of inclination would be removed when the Delta third stage fired—the rest would be removed by the AKM, which would be fired on the fourth apogee by command from the Andover Earth station.[55]

The satellite was about 28 inches in diameter and 23 inches high and weighed about 85 pounds in orbit. Mounted on the top of the satellite was the mast-like C-band (6 & 4 GHz) communications antenna. This antenna consisted of collinear slotted dipoles, which provided about 9 dB (8x) gain. The radiation pattern was a toroid (doughnut) symmetrical about the spin axis. A beam covering only Earth would provide about 19 dB (80x) gain, but this would require a three-axis stabilized antenna or an electronically de-spun phased array. The communications electronics consisted of redundant 6-GHz receivers cross-strapped to redundant 4-GHz 6-watt TWT transmitters. The communications system was capable of providing 240 duplex voice channels or one duplex television channel. The VHF (136 MHz) telemetry and command antennas were whips in a turnstile arrangement on the bottom of the satellite. Power was provided by 6,000 n-on-p silicon solar cells and two 21-cell nickel-cadmium (NiCd) batteries. The solar cells provided 45 watts in sunlight, but the batteries were needed to provide power during the spring and fall equinoctial periods when Earth eclipses the Sun (from the satellite's point of view) at midnight local time for up to seventy-two minutes at the peak of the two forty-four-day eclipse seasons.

The battery power was also needed to supply the peak power required by the attitude-control system solenoids. The communications payload was not operated during the eclipses. The propulsion system consisted of redundant hydrogen peroxide systems, each capable of providing in-plane (east-west, drift) and out-of-plane (north-south, inclination) Δv for orbit control, as well as torques for attitude control. This was a far more sophisticated spacecraft than the original 25-pound (11-kilogram) "commercial communications satellite" that the Hughes designers Harold Rosen, D. D. Williams, and Tom Hudspeth had envisioned in 1959, but it was clearly derived from the modified design they had developed in 1960.[56]

Early Bird was launched, on schedule (eighteen minutes after the opening of the launch window), at 7:48 P.M. EDT on Tuesday, April 6, 1965. Apogee was slightly lower than predicted, whereas perigee was somewhat higher. Overall, the Delta performed extremely well. The NASA STADAN stations tracked the satellite after launch—first, Johannesburg, South Africa, followed by Carnarvon and Woomera, Australia. Finally, twelve hours after launch, as *Early Bird* began its second orbit, Comsat's Andover station acquired the satellite signal. Just before the second apogee, Andover commanded the spacecraft to fire its axial thrusters in a pulsed maneuver to precess the satellite into the AKM firing attitude. On the fourth apogee, thrusters were fired to raise perigee. After a slight change to the attitude, the AKM was fired on the sixth apogee at 9:40 A.M. EDT on Friday, April 9, 1965. The final orbit had an inclination of 0.1 degree and an eccentricity of about 0.04. The satellite was over the equator at a longitude of about 33 degrees west drifting east at 0.2 degree per day. Hydrogen peroxide thrusters were used to maintain the satellite east-west position (semimajor axis) between 28 degrees and 38 degrees west longitude. Control of the north-south position (inclination) was not attempted due to the large amount of fuel required. Over time, the north-south latitude excursions (inclination) would grow by about 0.9 degree per year.[57]

Formal communications experiments began on April 10, but Reiger had performed a television "pre-test" while the satellite was still in geosynchronous transfer orbit. Commercial service was scheduled to begin on June 1. *Early Bird*'s 240-voice-channel capacity was almost equal to the 317-channel capacity of all the existing Atlantic cables—and much cheaper! Existing stations in the United States (Andover), the United Kingdom, France, Italy, and Germany would begin offering commercial service in the summer. Other stations were under construction in Canada, Spain, and around the world. Comsat was petitioning the FCC to allow it to build stations in the states of Washington and Hawaii in anticipation of Pacific Ocean service. An article in *U.S. News & World Report* emphasized transatlantic television as the most "visible" capability of the new satellite but also examined the economics of satellite communications. The latest-technology cable carried fewer channels and cost about ten times as much as the satellite. *Early Bird* and its Delta launch vehicle had cost Comsat only about $7 million—a small portion of its $200 million capitalization. The article quoted Comsat officials as stating that GEO had been picked for initial testing because of its simplicity and cheapness but that the decision on Comsat's "basic system" had still not been made—and would not be made until the end of the year. On May 2, 1965, a global television extravaganza was held linking Europe and America. On May 13 ABC filed with the FCC for permission to launch a television relay satellite—the first "domestic" communications satellite. Earlier in the year, before the launch of *Early Bird,* Hughes had suggested that the technology to build a television broadcast satellite was now available. Rosen, of Hughes, suggested that Arthur C. Clarke's 1945 dream of television broadcasts from space could be made a reality on the NASA ATS program. Comsat's response to ABC (it apparently ignored the Hughes suggestion as being directed to NASA) was that Comsat had a congressionally mandated monopoly on satellite communications but would be glad to provide a relay service for ABC. Estimates of AT&T revenues for relaying television on terrestrial circuits was about $50 million per year. The Comsat-Intelsat satellites would be dominated by telephone traffic for many years, but it was clear that television had captured the public's attention and had begun to create the "global village."[58]

In June AT&T filed with the FCC for permission to lease 100 voice channels from Comsat and also expressed interest in occasional television transmission. AT&T was still negotiating with the European PTT (Post, Telegraph, and Telephone) administrations but expected to have 36 links with the United Kingdom, 12 with Germany, 10 with France, and more with other countries. AT&T suggested that it was satisfied with Comsat's proposed rates of $4,200 per month for voice circuits and $2,400 for a twenty-minute telecast. Within two days RCA had filed for 30 circuits, and within a week ITT had filed for 41 circuits. Western Union International (WUI) later filed for 55 circuits. A total of 226 voice circuits out of 240 available had been requested. The official start of commercial services was to be June 27, but emergency permission to begin commercial services was received from the FCC after the failure on June 17–18 of one of the transatlantic cables. On June 23 the FCC granted Comsat the authority to begin operations, although the rates were not set. Revenues were to be held in a special fund until rates were finalized by the FCC. Voice circuits were allocated by the FCC to AT&T (75), ITT (10), WUI (10), RCA (10), and Canada (6). Allocations were limited to 120 circuits rather than 240 because European stations were not yet capable of handling the full traffic load. A long article in the August 2, 1965, *Wall Street Journal* bemoaned the laggardness of the Europeans. AT&T had managed to find partners for 60 of its circuits, but only one other circuit was in operation—an RCA circuit to Germany. AT&T paid Comsat $250,000 for its first month of satellite circuit use. The Europeans were making it clear that international telecommunications required two equal partners. A U.S.-dominated global satellite communications system would not be tolerated.[59]

At the May 11, 1965, shareholders meeting a new board was elected. There were no changes in the public directors at that time. One of the presidentially appointed directors, Clark Kerr of the University of California, was replaced by William W. Hagerty of Drexel Institute. A few months later Comsat announced that the chairman, Leo Welch, would be stepping down. In October the board of directors elected James McCormack as the new chairman. He replaced Leonard Marks, who had resigned in September, on the board. McCormack had degrees from West Point, Oxford, and MIT. He had retired from the Air Force as a major general and had subsequently served as the first president of the Institute for Defense Analyses and as vice-president of MIT. He brought technical and political expertise to the board, as well as the experience of managing large organizations. This board, with minor changes, was to serve Comsat for many years.[60]

Intelsat 2

In 1963 Pat Hyland, Hughes general manager, had suggested to Bob Gilruth, of NASA, that Syncom could provide the Apollo project with a high-quality, wideband, global communications system linking NASA's tracking stations in real time with the Mission Control Center. Without the satellite link, NASA was dependent on cable and radio links, which were unreliable and narrowband. Much data was recorded on magnetic tape and shipped to NASA. Apollo would have more data to transmit—and possibly even television. This idea was discussed for the next two years and included discussions of the use of *Early Bird* for this purpose. *Early Bird* had two disqualifying features. First, the *Early Bird* beam was "squinted" to cover only the Northern Hemisphere. Second, the receivers on *Early Bird* incorporated limiters, which made the system unsuitable for multiple-access use. Only a single pair of Earth stations could use the satellite at one time. Providentially, the Intermediate Syncom design could provide all of these capabilities. There may have been some interest in launching a NASA-owned and operated system, but the "dual-use" arguments between DoD and Comsat would have made it clear that this would be at least controversial and possibly illegal. Even assuming Comsat-owned satellites, many questions would have to be addressed regarding service providers and equipment owners.[61]

In June 1965 NASA requested that the manager of the NCS determine if Comsat could provide satellite communications services to the Apollo Network. On

July 26, 1965, NCS invited Comsat to quote prices and provide technical details on how it might provide these services. On September 30, 1965, Comsat filed an application with the FCC for authorization to build and launch four geosynchronous communications satellites to provide communications services for NASA. Two would be launched in late 1966—one to provide services over the Atlantic and one over the Pacific. On October 17 Comsat notified the FCC that it proposed to buy four satellites from Hughes, at a total cost of $11.7 million. The Apollo Network would include three Comsat stations in the United States (Andover, Maine; Brewster, Washington; Paumalu, Hawaii), a Cable & Wireless station on Ascension Island, three U.S. government shipboard stations, a station at Carnarvon, Australia, and a station on Grand Canary Island, Spain. By November, Intelsat had announced that it had approved a contract with Hughes for a satellite that would be placed over the Pacific Ocean. In the meantime, Comsat advised NASA that it would not be launching any more HS-303 *(Early Bird)* satellites using the thrust-augmented-Delta (TAD) but would be launching four HS-303A (Intelsat 2) satellites using the thrust-augmented-improved-Delta (TAID). NASA assumed a total liability (money it would pay if it did not use the system) of $10.5 million for the entire network. At a time when only one-quarter of the *Early Bird* capacity was being used, Comsat had acquired a major new customer.[62]

Intelsat 3 and 4

Although two generations of Comsat-Intelsat satellites had been ordered, no decision on the "basic system" had yet been made. *Early Bird (Intelsat 1)* was both an experimental and an operational system. The Intelsat 2 satellites formed a special-purpose system launched to satisfy NASA's needs. At the same time that Hughes had received the *Early Bird* GEO contract, two MEO study contracts had been awarded to an AT&T-RCA team and a TRW-ITT team. The assumption at the time was that an MEO experimental-operational system would follow *Early Bird* sometime in 1966. The pace was too swift to allow this kind of experimentation—and the Telstar and Relay experiments were assumed to have proven the MEO case. It was time to choose. Hughes, the builder of *Early Bird* and Intelsat 2 as well as the partner in Comsat's bid to win the DoD communications satellite contract, was proposing an "advanced Early Bird," the geosynchronous HS-304. AT&T and RCA, builders of Telstar and Relay and representing one-third of the Comsat shareholders and the dominant telecommunications entity (AT&T), were proposing a system of 18 satellites in 6,000-mile (11,000-kilometer) "random" polar orbits. TRW and ITT were proposing a controlled or phased system of 12 satellites in similar orbits. In May Charyk announced that Comsat was no longer considering the random approach. It was now convinced (possibly by the *Early Bird* experience) that controlling GEO satellites was not as difficult as earlier assumed. Charyk noted that a satellite that could operate at both MEO and GEO would considerably reduce design costs.[63]

On August 17, 1965, Comsat issued, on behalf of the Interim Communications Satellite Committee (ICSC), an RFP for an "advanced satellite." The satellite would be for use either in a GEO or in a phased system at altitudes between 6,000 and 12,000 miles (MEO). Preference was to be given to a design capable of using *both* MEO and GEO orbits. Proposals would be opened on October 25, 1965. The manufacturers would have seventy days to respond. The RFP specified a capacity of 1,000 two-way voice circuits, a five-year lifetime, a weight of approximately 240 pounds, two repeaters, a directional antenna, and multiple-access capability. On December 16, 1965, Comsat announced that it was negotiating with TRW for at least six Intelsat 3 satellites at a cost of approximately $20 million. RCA and Hughes had also bid for the contract. TRW had offered a design capable of operating at both MEO and GEO altitudes. Comsat still had not decided which of these orbits to use. After almost three years of existence and three generations of satellites ordered, Comsat had over $180 million of its initial $200 million capitalization and only 75 of 240 voice circuits on *Early Bird* in use.[64]

On December 29, 1965, Comsat issued, on behalf of Intelsat, an RFP for design studies leading to its fourth-generation satellites. These would have a capacity of 6,000 voice circuits (or ten TV channels), would have a five-year lifetime, solar power was preferred but nuclear power could be proposed, and would weigh less than 2,300 pounds (one ton). The satellites would be geosynchronous. Comsat had finally chosen a system![65]

On the last day of 1965, Comsat released its first quarterly report. Comsat listed revenues of $966,000 from *Early Bird* operations. Given that these revenues were gathered during the first quarter of the life of an investment in excess of $7 million, those figures were quite disappointing. On the other hand, *Early Bird* was primarily an experiment—and it had been an extremely successful experiment. Comsat still had almost $188 million in cash. It had contracted to provide Atlantic and Pacific Ocean service to support the Apollo program. The FCC had granted Comsat sole responsibility for building and operating communications satellite Earth stations in the United States. Comsat had contracted with Sylvania to build two new Earth stations, in Hawaii and Washington. It was also in the process of consolidating all of its operations into one building, at L'Enfant Plaza in downtown Washington, D.C. Last but not least, Intelsat, on Comsat's recommendation, had chosen GEO for its "basic system." Arthur C. Clarke could be proud, but so could John R. Pierce, Harold A. Rosen, Donald D. Williams, Tom Hudspeth, Sid Metzger, Siegfried Reiger, Leonard Jaffe, and the hundreds of other engineers who had helped to make commercial satellite communications a reality. Political forces may have determined the form of the final system, and economic forces had provided most of the impetus, but none of these forces would have produced the global communications satellite system without the efforts of the engineers and technicians who had envisioned, developed, demonstrated, and deployed the "billion-dollar technology."[66]

7. Outcomes

Billion-Dollar Industry

The attempt to have NASA re-enter communications work with their efficiency and at taxpayer expense is a real issue and we should not let people misread the record to support that folly. —E. F. O'Neill (AT&T), August 17, 1977

It is fascinating that AT&T engineers, especially John Pierce, seemed very willing—at least after the fact—to give Hughes Aircraft Company and Harold Rosen credit for inventing a "good" synchronous satellite. In 1966 Pierce seems to have felt that Rosen and his associates deserved their claim to fame but that NASA claimed far more than it deserved.[1] Thirty years later, Pierce still felt that only AT&T and Hughes had made significant contributions to the origins of communications satellite technology. Admitting that AT&T had its own faults, Pierce remained critical of the government's anti-AT&T bias—especially the resulting exclusion of AT&T from global satellite communications and the failure to credit the technological advances made by AT&T engineers.[2]

Arthur C. Clarke is credited with the "discovery" of GEO (geosynchronous Earth orbit), often called the Clarke orbit or Clarke belt. Clarke himself saw his role as much smaller and in many ways premature. He credited Pierce and Rosen as the true "fathers" of communications satellites and considered himself the "godfather." This opinion was seconded by the National Academy of Engineering, which awarded Pierce and Rosen the Charles Stark Draper Prize in 1995.

Also in 1995 were two conferences—one held at George Washington University and one at NASA Headquarters—celebrating Rosen and Pierce's

Harold A. Rosen (second from left) and John R. Pierce (second from right) receiving the Draper Prize from the National Academy of Engineering in 1995. (Courtesy NAE)

Draper Prize, the thirty-fifth anniversary of *Echo I*, and the thirtieth anniversary of *Early Bird*.[3] John Rubel (former director of defense research and engineering) was the first speaker at the George Washington University conference. He discussed his long friendship with Rosen and other Hughes employees. In 1960, after leaving Hughes to work in government, Rubel was agonizing over the continuing problems with the three-axis geosynchronous Advent communications satellite program. A friend, Gordon Murphy, invited Rubel to visit Hughes and see what Rosen and his team had accomplished. Rubel was impressed and began to push for Syncom with Robert C. Seamans Jr. at NASA and with his own colleagues at the Department of Defense (DoD). According to Rubel, one of the naysayers was Hugh Dryden, not because of Syncom itself but because he feared congressional reaction.

Harold A. Rosen was the second speaker at this conference. He retold the story of the beginnings of Syncom, emphasizing that the design was based on what could be done, not on some organization's unrealistic requirements. According to Rosen, Rubel made it all happen. Rosen and other Hughes employees had made the rounds, with no success. Rubel got NASA and DoD to cosponsor the *Syncom* program.

In his talk, Rubel also stated that he and Fred Dutton, a White House staffer, were principal contributors to the Communications Satellite Act of 1962, perhaps more so than Dr. Edward Welsh, of the National Aeronautics and Space Council. At the same meeting, John A. Johnson, NASA general counsel in the early 1960s and later the general counsel for Communications Satellite (Comsat) Corporation, claimed that the Kerr bill had been written at NASA. Johnson also claimed that both NASA and the Federal Communications Commission (FCC) had thought the communications companies should own and operate the satellite communications system but that the White House had apparently objected.

Leonard Jaffe also spoke at George Washington University. In the late summer of 1958, he had been asked by Abe Silverstein, his old boss, to come to NACA (National Advisory Committee on Aeronautics) Headquarters from the NACA Lewis Research Lab and, once there, to attend a meeting on satellite communications at the Pentagon. NASA did not yet exist, nor did Jaffe have a firm position at NACA Headquarters. The subject of the Pentagon meeting was defense satellite communications generally and the geosynchronous Advent program specifically. According to Jaffe, Pierce had problems at this and other meetings with the lack of realism in the Advent program. Jaffe concluded his talk with the observation that NASA and DoD had been the drivers in the development of satellite communications and had brought industry into the endeavor.

Sid Metzger, who had been the driving force at RCA during the Relay program, spoke at the George Washington University conference and also attended the meeting at NASA Headquarters. RCA's interest in satellites had begun with its participation in the RAND study of reconnaissance satellites in 1954. RCA had bid a spinning reconnaissance satellite in 1955 and had lost to the Lockheed three-axis design. Later, DoD officials had suggested that the RCA reconnaissance satellite design might make an excellent weather satellite (this became TIROS). The RCA Relay design benefited from these experiences. Metzger went to Comsat in 1963 and helped organize the company. In December 1963, over the objections of many of the Intelsat partners, Comsat decided to award a contract to Hughes to build the GEO *Early Bird* and to use the 6-GHz and 4-GHz frequencies. In later discussions and in correspondence over the next few years, Metzger affirmed his belief that only government could have brought satellite communications into being.[4]

Policy Choices through 1965

The historical events related here should make it clear that the early development of communications satellite technology was performed by AT&T and Hughes Aircraft Company using internal funding. These companies were willing to spend their own money because they expected to make substantial profits from communications satellite technology. AT&T provided the complete funding for Telstar. Hughes provided Syncom funding until it had a prototype to show potential investors; it might have provided more funding if the government had not stepped in.

History also shows that rocket development, although initiated by "space cadets," was completed by the military for purposes of national defense. Even the peaceful sounding-rocket technology that led to Vanguard was funded by the military. Although neither Hughes nor AT&T received funding for space projects before their development of Telstar and Relay, both companies had been major defense contractors. This defense funding was probably trivial in the case of AT&T, but it was critical to the growth of Hughes from a "hobby-shop" aircraft manufacturer to a major electronics company.

Market forces produced the transoceanic communications services demand that communications satellites could satisfy. Even before the launch of the first communications satellites, industry saw potential profit and was more than willing to invest tens of millions of dollars (Hughes, General Electric, Lockheed, ITT, and others) and even hundreds of millions of dollars (AT&T). NASA spent over $300 million between 1959 and 1976 on communications satellites.[5] During this same period, industry spent over $800 million[6] of its own money: $100 million on R&D; over $500 million[7] on the first four generations of international communications satellites; and over $200 million[8] on domestic communications satellites. Most of the NASA spending was on the ATS program, rather than on the Relay and Syncom programs, which were relatively inexpensive. Industry could have and would have implemented a global satellite communications system—albeit a system dominated by AT&T and MEO (medium Earth orbit) satellites. Why, then, did the government intervene?

Satellite communications was part of the space race, and the space race was part of the cold war. The cold war had taken a turn for the worse since the mid-1950s—at least from the U.S. point of view. The Communist challenge in Korea had been met; U.S. atomic bombs were bigger than Soviet bombs; and even the explosion of a Soviet thermonuclear device suggested only parity, not superiority. This began to change in 1956, when the United States did not support the Polish and Hungarian uprisings. Neither did the United States support its allies during the 1956 Suez crisis. In 1957 the launch of *Sputnik 1* suggested Soviet superiority, not parity. In April 1961 the Bay of Pigs debacle and the orbital flight by the Soviet Yuri Gagarin were challenges that had to be met. President Dwight Eisenhower was reluctant to enter the space race in 1957, but the rest of the U.S. government was more than willing to pick up the gauntlet. The cold war challenge would be met in space. President John F. Kennedy's May 25, 1961, speech only confirmed what everyone already knew: the space race was America's opportunity to regain its status as the world's dominant economic, technological, and military power. What might have started out as a commercial venture was quickly seen by government—NASA, the White House, and Congress—as a means of displaying U.S. superiority in space technology to the world and to the voters at home.

Antitrust issues—perhaps more accurately, anti-AT&T issues—were also involved. AT&T's antitrust problems went back to early in the century, but the late 1950s saw the beginning of a regulatory turn against AT&T by the FCC as well as by the antitrust forces. The 1956 Consent Decree required AT&T to license, for free or at a reasonable cost, all of its patents. The decree also required AT&T to stay out of all businesses other than the telecommunications common-carrier business. At the same time, the FCC was allowing manufacturers of various new devices to connect them to Bell telephones. The FCC also was arguing that there were enough frequencies in the microwave bands ("above 890") to satisfy AT&T's telephony needs and still allow for other uses. The "above 890" filings were prominent in much of the congressional testimony in 1961 and earlier. AT&T and its employees behaved in an arrogant manner. One result was that AT&T was at least resented, and often hated, by competitors and regulators. This clearly affected the development of satellite communications policy.

The government thus had two reasons for intervening in the development of satellite communications: (1) winning the cold war, and (2) "regulating" (punishing) AT&T. The cold war in space seems the most important, but "regulating" AT&T was also important. What is not clear is whether this "regulation" was due to objective antitrust concerns or simply to a dislike of AT&T. During the discussions in the National Aeronautics and Space Council, the Department of State often urged government ownership of satellite communications to facilitate sharing this capability with the Third World. This Third World was often demonstrating against "U.S. imperialism" at the time. At least some members of both the First World and the Third World saw Comsat as an example of "U.S. imperialism"—a U.S. government monopoly. Intelsat, at least in part, was a counter to the Comsat monopoly.

Through 1965, several communications satellites had been launched, all by the United States and all—partially excepting *Early Bird*—as experimental satellites. As has been shown in the previous chapters, *Early Bird*, AT&T's Telstar, and the early Hughes Syncom were funded by industry. Although Echo was a government program, the telecommunications experiments were partially funded by AT&T. In the following thirty years government funding disappeared, except for military satellite procurement and the NASA ATS—and later ACTS—program.

The major outcomes of government intervention in the satellite communications market through 1965 were (1) forcing AT&T out of the business, except as a minority Comsat shareholder, (2) establishing Comsat/Intelsat as the single provider of global satellite communications, (3) positioning Hughes as the leading commercial satellite manufacturer, (4) establishing GEO as the preferred orbit, and (5) positioning U.S. companies as the leading manufacturers of satellites and launch vehicles. This last outcome was a given. The other outcomes might very well have occurred without government intervention with one very interesting exception: the forcing of AT&T out of the business.

Outcome #1: AT&T and Satellite Communications

AT&T had developed most of the basic technologies for satellite communications: transistors, solar cells, traveling-wave tubes, and more. AT&T had spent more of its own money than had any other nonmilitary organization. AT&T had also made international arrangements for its transoceanic submarine telephone cables—especially with the British Post Office. AT&T was the obvious choice to launch and operate an international communications satellite system, but it was not to be.

If AT&T had stayed in the satellite communications business, it would have built its own MEO satellites, and the exploitation of GEO might have been delayed. The monopsony power of AT&T was a significant barrier to the establishment of a GEO satellite system, but it was never clear that alleviating this problem was the government's intention. Most of the outcomes of government intervention were positive, but they were not necessarily the intended outcomes—nor, in the case of AT&T, were they fair.

AT&T was first blocked, in late 1960, by NASA's refusal to provide launch services. This seems unfair, but T. Keith Glennan, the first NASA administrator, was very much in favor of AT&T's plans, and the argument that launch vehicles were in short supply had a certain validity. Other Thor variants had been launched thirteen times—with five failures—before the first Thor-Delta. The new Thor-Delta had been used only twice—and had failed once. The often-made statement that launches were only 50 percent successful was a reasonable approximation of reality in 1960. There were over twenty successful Delta launches after the first Delta failure—a 95 percent success rate—but this could not have been foreseen. Somewhat stranger was NASA's refusal to take AT&T's offer to build Relay for $1. The government had clearly decided to control satellite communications development for its own purposes. Why else pay for R&D when industry was willing and able to take on the task? The patent issues are somewhat similar in nature. NASA claimed all patents for itself: those of AT&T and those of Hughes.

The politics surrounding the Communications Satellite Act of 1962 have been chronicled many times, but those chronicles leave out the positive contributions of AT&T. Those chronicles also leave out the many issues raised at the time—some publicly, some privately. There was significant anti-AT&T bias throughout the process. It is not clear what the intent of Congress and the White House really was. There was, however, a very clear bias among many for a government system. There was also a bias toward letting industry pay its own way. The final text of the Communications Satellite Act of 1962 seems to have been an attempt to have it both ways: creating Comsat as a public-private organization. There may have been reasons for excluding AT&T, but the result does seem unfair.

After 1965 AT&T was in and out of the satellite business. It sold off its Comsat stock, built and sold off Earth stations, but then reentered the satellite communications business in 1976 with the launch of Comstar in partnership with Comsat. This was not a particularly successful venture, since both AT&T and Comsat were seen as monopolies and had certain restrictions placed on Comstar. In the early 1980s AT&T developed its own system, without Comsat. This was called Telstar after the original satellite, but the Telstar system was eventually sold to Loral. The 1982–84 divestiture of AT&T did not help matters. The AT&T of the late twentieth and early twenty-first centuries was no longer a player in the satellite communications business.

Outcome #2: Comsat and Intelsat

As it became obvious that a U.S. government-sponsored organization, Comsat, would be given a U.S. monopoly of satellite communications, the European PTT (Post, Telegraph, and Telephone) administrations became concerned that they would be squeezed out of the picture. The Communications Satellite Act was passed in 1962, Comsat was formed in 1963, and Comsat went public in 1964. Intelsat was formed in 1964 to own and operate the global communications satellite system. Comsat would have a U.S. monopoly, but the PTTs would retain their national monopolies and a proportional ownership of Intelsat.

Because Intelsat operated on the one telephone call, one vote principle (a diplomatic triumph itself) and U.S. traffic dominated the original satellites, Comsat was able to control the Intelsat decision-making process for several years. Somewhat surprising to many is the fact that *Early Bird*, launched in 1965, was an experiment. Comsat had not yet decided on the form of the basic global communications satellite system, but a GEO experiment using commercial frequencies would be cheap. The *Early Bird* satellite performed well, but demand was lower than expected. NASA offered to be an anchor tenant on the

Table 7.1
Communications Satellite Launch Rate, 1966–1995 (Operators)

	Military	International	United States	Europe	Japan	Other	Total
1966–1975	44	20	9	2	1	3	79
1976–1985	38	20	32	12	10	23	135
1986–1995	24	20	36	30	18	32	160

Intelsat 2 series: a certain level of NASA revenues was guaranteed to Comsat-Intelsat in exchange for ensuring that a global network was available for Apollo support. The Intelsat 3 series was designed to be launched into either MEO or GEO. The first series of the basic global system that represented a commitment to GEO was Intelsat 4. There was some international participation in the development of both Intelsat 3 and Intelsat 4.

Comsat and Intelsat evolved substantially over time. Satellite operations were initially governed by an interim agreement (1964) that allowed Comsat considerable latitude. In 1973 a more formal arrangement entered into force, under which Comsat became the Intelsat Management Systems Contractor until February 1979. The Intelsat 3 series, first launched in 1968, was the first to require international participation in manufacturing, the first to be approved by the entire Intelsat community, and the first to service all three ocean regions: Atlantic (AOR), Pacific (POR), and Indian (IOR). The Intelsat 4 series, manufactured by Hughes, established a standard C-band (6 GHz up, 4 GHz down) payload configuration that persists to this day on most commercial communications satellites: twelve (twenty-four with polarization reuse) 36-MHz transponders.[9] During the period 1966–75, Comsat-Intelsat launched twenty satellites. These twenty satellites dominated transoceanic communications and in the process made the world a smaller place. It is interesting to note (see Table 7.1) that the Intelsat launch rate remained at about two satellites per year for thirty years.

The original intent of the Communications Satellite Act of 1962 was to form a single global communications satellite system (preferably under U.S. control). Comsat used this intent to delay the military IDCSP (Interim Defense Communications Satellite Program) and to delay regional-domestic systems. Comsat was aided and abetted by NASA and the rest of the U.S. government. In 1972 Telesat Canada launched the first GEO domestic communications satellite service on *Anik-A1*. The rocket was a Delta; the satellite was a Hughes HS-333. In 1970 the Nixon administration approved an "open skies" program under which private companies could launch their own systems. Western Union launched the first U.S. system, *Westar 1,* in 1974. RCA followed quickly in 1975 and Comsat-AT&T in 1976. Indonesia launched its own domestic system, Palapa, in 1976. These were the wave of the future.

After 1975, non-Intelsat satellites were more common than Intelsat satellites. After 1985, non-Intelsat, non-U.S. satellites continued to grow in number. These new satellites were mainly used for video distribution rather than telephony. The next-largest market was private networks using very small aperture terminals (VSATs). In 1984 the United States passed "separate systems" rules that allowed other operators to compete with Comsat-Intelsat for the international switched telephony market. Comsat and Intelsat were supplying a dying market. It is not surprising that Comsat was sold in 2000. But it is surprising that Comsat was sold to Lockheed Martin. Intelsat privatized in 2001. It will be interesting to see what form the new Intelsat takes. As for Comsat, the chief executive officer of the merged company, Lockheed Martin Global Telecommunications (LMGT), stated many times that the company would not be a satellite communications company.[10] In late 2001 Lockheed announced the dissolution of LMGT.

Outcome #3: Hughes Aircraft Company

Hughes had a better idea in 1959–60: a simple, lightweight, geosynchronous communications satellite. The Rosen team's first thought was for Hughes to launch and operate the satellite itself. The second idea was to team with a telecommunications company. Hughes General Manager Pat Hyland's idea turned out to be prescient: get the government to support initial development without giving up patent rights, and then sell to everyone. Hughes made the first commercial sale of a communications satellite, *Early Bird*, to Comsat in December 1963. It also sold the Intelsat 2 and Intelsat 4 series to Comsat-Intelsat. Hughes did not bid on the Intelsat 3 series because it objected to the idea that the satellite must be capable of MEO operation. Hughes built the first five ATS program satellites in the late 1960s. By the end of 1975, Hughes had built twenty-eight GEO communications satellites: fourteen for Comsat-Intelsat, nine for the U.S. government, and five for domestic communications satellite operators.

After completing the Intelsat 4A series with the launch of *Intelsat 4A F-6* in 1978, Hughes backed away from Intelsat procurements. Hughes wanted to build production line satellites for regional and domestic operators who were not as picky as Intelsat. The HS-333 series was a small, lightweight satellite with twelve low-power transponders and a single, relatively large antenna. Eight HS-333s were built. The first ANIK (Telesat Canada), Westar (Western Union), and Palapa (Indonesia) satellites were all part of this series. In the mid-1970s, Hughes faced its first real challenges: Ford Aerospace won the Intelsat 5

Tom Hudspeth (left) and Harold A. Rosen with a Syncom prototype in front of a full-scale mockup of HS-601 in 1993. (Courtesy Hughes/BSS)

procurement; and RCA Astro-Electronics built a medium-power, 24-transponder, three-axis stabilized satellite that it sold to domestic U.S. customers. Hughes built a spinner with a similar payload (Intelsat 4A and Comstar), but it was heavier and more expensive. The Hughes reaction to the challenge by Ford and RCA was to build a more-sophisticated lightweight spinner with medium power: the HS-376. The first HS-376, *SBS-1*, was launched in 1980. The HS-

333 had been the original "entry-level" satellite for the domestic and regional communications satellite operators; the HS-376 was to be the new "entry-level" satellite. More than fifty HS-376s would eventually be ordered. By the late-1980s, many of the original "entry-level" operators were looking for a higher-powered satellite. The spin-stabilized HS-376 was limited to about 1,000 watts but was competing against 2,000–5,000-watt, three-axis stabilized satellites built by Ford and RCA. Hughes responded with the three-axis stabilized HS-601. The first HS-601, *Optus B1*, was launched in 1992. More than seventy have been ordered since that year.

Except for Leasesat and UHF Follow-On (UFO), both of which were procured under (almost) commercial conditions, Hughes has avoided the geosynchronous military communications satellite market. Similarly, Hughes has backed away from the Intelsat market. Since the mid-1970s, domestic and regional satellite systems have seen most of the growth, and Hughes has consistently captured about 40 percent of this market. Hughes entered the ranks of communications satellite operators in 1983 with the launch of *Galaxy I*. In 1997 Hughes merged its own communications satellite operation with PanAmSat to create the world's largest communications satellite operator. In 1993 Hughes became the first operator of the U.S. Broadcast Satellite Service (direct-to-home television) with the launch of *DirecTV 1*. As the communications satellite manufacturing market has matured, Hughes has become the biggest player in the communications satellite operations market.

For most of its existence, Hughes Aircraft Company was owned by the Hughes Medical Institute. After the death of Howard Hughes, the Institute—a grant-giving philanthropic institution—was forced to spend its money or lose its nonprofit status. Eventually Hughes was sold to General Motors (GM). In the late 1990s GM began selling off the pieces of Hughes. The original radar group was sold to Raytheon. Hughes Space and Communications—premier manufacturer of communications satellites—was sold to Boeing.

Outcome #4: GEO

The Telstar and Relay satellites were intended to operate in circular, inclined orbits, at an altitude of approximately 10,000 kilometers (MEO). Existing launch vehicles could not even deliver circular MEO orbits. GEO was seen as too difficult because the orbital velocity (Δv) that was needed to reach GEO was far beyond the capability of existing launch vehicles and because GEO satellites needed complex orbit- and attitude-control systems. Most analysts viewed GEO as a better orbit for satellite communications but perceived the problems associated with it as disqualifying. Syncom changed all this. Except for the IDCSP

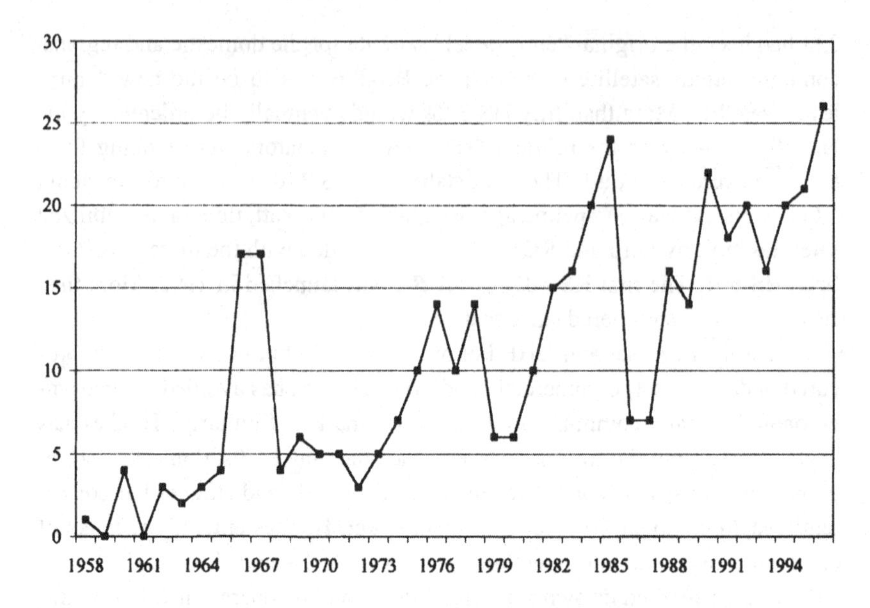

Figure 7.1. Communications Satellite Launch Rate, 1958–1996.

satellites, which were in a drifting-near-GEO orbit, all operational communications satellites launched after *Syncom 3* were GEO—until the late 1990s.

In the 1990s several LEO (low Earth orbit) systems were launched, and several MEO systems were proposed. As of late 2000, two out of three LEO systems launched were in bankruptcy, and the third was not earning a profit. None of the MEO systems had been launched. Interestingly Iridium, one of the bankrupt LEO systems, was sold off for a small fraction of its cost as the result of the bankruptcy. The "new" Iridium has a small but apparently thriving business.

Outcome #5: U.S. Dominance of Manufactures

Figure 7.1 depicts the launch rate of communications satellites from 1958 to 1996. There are many peaks and valleys, but a general upward trend is obvious—especially after discounting the 1966–67 anomaly due to the launch of the IDCSP satellites. The peaks and valleys result from a combination of replacement cycles—satellites grow old, run out of fuel, and are replaced—and the birth of new satellite markets. The drop in the mid-1980s is somewhat more complicated, since it is also associated with the *Challenger* disaster in January 1986. Just before this accident, the communications satellite launch rate had been around twenty launches per year. In the last few years of the twentieth century, the launch rates averaged closer to thirty per year. Although dividing

Table 7.2
Communications Satellite Launch Rate, 1966–1995 (Manufacturers)

	United States	Europe	Japan	Other	Total
1966–1975	77	2	0	0	79
1976–1985	116	11	2	6	135
1986–1995	113	34	4	9	160

Table 7.3
Communications Satellite Launch Rate, 1966–1995 (Launchers)

	United States	Europe	Japan	China	Total
1966–1975	78	0	1	0	79
1976–1985	107	18	9	1	135
1986–1995	58	84	7	11	160

the thirty-year span from 1966 to 1995 into ten-year periods is somewhat arbitrary, it is interesting to examine the changes in the nationality of operators, manufacturers, and launchers during each of these periods.

In Table 7.1, the launch rate in each of the three ten-year periods is tabulated by the nationality of the *satellite operator*. Compared with the dozen or so communications satellites launched through 1965, the growth is astounding. However, several trends are interesting: military launches per decade decreased, international launches (e.g., Intelsat) held constant, and regional and domestic satellite launches increased.

In Table 7.2, the launch rate in each of the three ten-year periods is tabulated by the nationality of the *satellite manufacturer*. Although U.S. dominance decreased over the years—and has further decreased since 1995—U.S. manufacturers still captured about 70 percent of the total between 1986 and 1995.

In Table 7.3, the launch rate in each of the three ten-year periods is tabulated by the nationality of the *launch vehicle manufacturer*. A fascinating statistic appears in this table: since the *Challenger* disaster in 1986, Europe has dominated the commercial launch industry with its Ariane rocket. This can be traced to several events: (1) the refusal of the United States to launch the Franco-German *Symphonie* satellite, except under onerous conditions, in the 1970s; (2) the emphasis on shuttle-only, which resulted in the near-elimination of U.S. expendable launch vehicles; and (3) the *Challenger* disaster. All of these had more to do with U.S. government policy than with U.S. technology or U.S. competitiveness in the marketplace.

While recognizing that these ten-year periods are arbitrary, we can still describe the first period as characterized by U.S. dominance and the rise of Intelsat. From 1966 to 1975, 79 communications satellites were launched. All but six were operated by U.S. organizations, were manufactured by U.S. companies, and were launched on U.S. rockets. Of the 79 satellites launched, 77 (97 percent) were manufactured in the United States, and 78 (99 percent) were launched by U.S. rockets. All but 15 were military and Intelsat satellites. Of these 79 satellites, one-third were manufactured by Hughes, and one-third were launched on Delta rockets. When military satellites are excluded, Hughes and Delta dominated the market.

The second period was characterized by the advent of domestic satellites. From 1976 to 1985, 135 communications satellites were launched. U.S. manufacturers (116) and U.S. rockets (107) still dominated in this period, but operators became much more diverse: military satellites and international satellites no longer dominated the market. Of the satellites launched during this period, only 58 were military or international. Of the other 77 satellites, U.S. domestic satellites, used extensively for cable television program distribution, were the single largest component. In spite of warnings in the 1970s and later that the U.S. control of satellite manufacture might be challenged, U.S. satellite manufacturers—especially Hughes—still dominated the market.

The third period, of crowded skies, runs from 1986 to 1995. In this period, 160 communications satellites were launched. U.S. manufacturers (113) still dominated satellite manufacturing, but U.S. rockets (58) had little more than one-third of the market. During this period the U.S. control of satellite manufacturing continued, but the aging U.S. ELVs lost ground to Ariane. Of 160 satellites launched in this period, 113 (71 percent) were manufactured in the United States, but only 58 (36 percent) were launched on U.S. rockets.[11]

NASA

NASA had initially planned to continue testing and evaluating each of the three technologies associated with Echo (LEO passive), Relay (MEO active), and Syncom (GEO active). *Echo 2* was eventually launched in 1964, but neither the advanced Relay satellite nor the advanced Syncom satellite was flown—to a large extent because Congress did not want to fund commercial projects for a monopoly (Comsat), not even a congressionally mandated monopoly. After the passage of the Communications Satellite Act of 1962, and especially after the formation of Comsat in 1963, the U.S. Congress was unwilling to fund NASA research in communications satellites. The rationale was that this area was now a private matter for Comsat to manage. By broadening its scope to "applications," NASA was able to continue communications satellite research under the aegis of the ATS program. This program launched six satellites through 1974. All except *ATS-6* were launched on the more powerful Atlas rocket and tested new technologies. For a variety of reasons, the direct transfer of ATS technology to commercial satellites was rare. Perhaps the most significant reason was that industry was already proceeding with commercial programs—including R&D. These programs paralleled the technology development of the ATS program. However, it should be noted that the ATS program, even if it did not lead the way, clearly established certain technologies as practical and showed that others were not ripe for utilization in commercial communications satellites.

The ATS program came to an end with the launch of *ATS-6* in 1974. In 1973 Congress had directed NASA to stop performing communications satellite R&D that could benefit only Comsat—a private firm. Many observers, especially within the NASA community, felt this was a mistake. Their fear was that communication satellite R&D funding by foreign governments would allow the satellite market to be taken from the United States.

Complicating this story are the different motivations of NASA, Comsat, and the more purely commercial satellite communications organizations. NASA was politically motivated at the top; its real purpose was to beat the Soviets—at everything. Sometimes this benefited the development of communications satellites, but often it did not. Comsat was a public corporation, created by government but otherwise commercial. It had monopoly powers and was generally regarded as the chosen instrument of the U.S. government. Comsat and Intelsat often used their political power against the commercial companies.

After the last ATS launch in 1974, NASA interest groups conducted extensive lobbying, arguing that foreign investment in communications satellite R&D was eroding the U.S. share of the market. The NASA experimental Advanced Communications Technology Satellite (ACTS) was finally approved (and reapproved) in the mid-1980s. Surprisingly, no serious effort was made to develop a new, expendable launch vehicle. During this period, the U.S. share of the launch vehicle market began to decline.

NASA reemerged as a player in the communications satellite arena with the launch of ACTS in 1993. Two new technologies appear to have surfaced coincident with the launch of ACTS: (1) onboard communications processing, and (2) use of Ka-band frequencies (30/20 GHz). It should be noted that major investment in onboard processing by Motorola-Iridium (and Ford Aerospace) predates the launch of ACTS. Interest in Ka-band is more complex. The very high rain fade of Ka-band made it an unlikely technology for decades. This disadvantage persists, but a new application, packet data (especially Internet data), may have overcome this disadvantage. The primary effect of rain fade is to lower the availability of Ka-band transmission links from the traditional 99.99 percent (i.e., one hour of outage per year) of commercial telephony to perhaps 99 percent (eighty hours of outage per year), which may be acceptable to consumers but not to businesses. None of the new Ka-band services have actually been launched, but interest is high. Is this interest due to changes in the market, or is it due to ACTS demonstrations?

Last Thoughts

In the preface, the purpose of this book was stated to be the challenging of the conventional wisdom that the government developed satellite communications because industry could not. The preceding chapters have shown the conventional wisdom to be wrong. Industry developed satellite communications because a market for transoceanic communications already existed and because companies—at least AT&T and Hughes—expected to profit from satellite communications. The government intervened in satellite communications and affected outcomes. Although much evidence remains to be examined, it seems fairly clear that government intervention was an example of "technology for policy." The government did not intervene to advance the technology but rather to advance certain elements of domestic and international policy.

The critical event was the launch of *Sputnik 1*. Although rocket development was common knowledge, the actual orbiting of a man-made object galvanized many of the players—especially John Pierce of AT&T. For Hughes Aircraft Company, there was another significant event: the cancellation of the F-108 fighter. Hughes wanted to diversify into space. But *Sputnik 1* did more than encourage the "space cadets"; it turned the U.S. space program into a competition with the Soviet Union. This may have been the main reason that the government intervened in satellite communications.

Satellite communications also provides examples of technology "push" and market "pull." The transoceanic telephony market was growing by 20 percent every year. AT&T—and to a lesser extent, Hughes and other telecommunications companies—saw an opportunity to profit from any improved method of transoceanic communications. Rocket technology made it all possible, and the Hughes invention of a practical, lightweight, simple geosynchronous satellite made satellite communications inexpensive. Although market "pull" seems more robust than technology "push"—especially since markets fund technology development—it seems unlikely that rockets would have been developed commercially without government sponsorship.

The economic problems of monopsony, specifically the tendency of AT&T

to buy exclusively from its subsidiary Western Electric, seem never to have been addressed directly by the government. In spite of the 1956 Consent Decree and continued discussion of this problem, monopsony never seems to have entered the discussions of satellite communications policy.

Finally, an unanswered question revolves around the perceived need by many in government—then and now—for a communications satellite *policy*. Is government policy required for the birth of any new industry? Is a specific "policy for technology" required for each new technological market advance? Government intervention in the United States seems to have affected communications satellite outcomes, but the government did not create this technology—nor did it create the market for this technology. Satellite communications was and is a unique space market. This market was not created by government. However, the enabling technology—rocketry—was clearly a government development. Satellites themselves were initially developed by industry in the United States—generally, but not exclusively, using internal funds. Does the government need to take credit for every economic or technological advance? The development of the communications satellite market suggests that governments are unlikely to develop completely new space markets. They may advance or retard these markets—especially with the development of enabling technologies and favorable climates for innovation—but other forces will establish the markets.

The above notwithstanding, a "policy for technology" is important. As discussed in chapter 1, the United States has always had some form of "policy for technology." These policies have served the country well, as we have been the masters of innovation for well over a century. Most of these policies, at least before World War II, attempted to facilitate invention and innovation. Since World War II, there has been an expectation that invention and innovation can be "forced" by the government. This may be the way to make pâté de foie gras, but it is not an efficient way to develop technology. In spite of much rhetoric, satellite communications remains the only profitable commercial space business.

Appendix 1

Orbits

During the early 1960s, the choice of an orbit for communications satellites was the single most important decision debated by designers. Orbit choice seemed to drive system design constraints. The higher the orbit, the lighter the satellite had to be for a given launch vehicle (rocket). The higher the orbit, the more powerful the satellite transmitter would have to be or the more directional the satellite antenna would have to be. The lower the orbit, the smaller would be the satellite footprint and the greater would be the need for multiple satellites—since each would be visible for only a few minutes—and the greater would be the need for complicated tracking systems.

Orbit choice was further complicated by the discovery of the van Allen belts—belts of high-intensity radiation that would damage a satellite that spent too much time in them. There are two rather broad and diffuse "belts": the lower belt extends from approximately 1,500 kilometers above Earth's surface to approximately 5,000 kilometers; the outer belt extends from approximately 13,000 kilometers above Earth's surface to approximately 20,000 kilometers. These belts constrain satellites to orbits below the lowest belt (LEO), between the belts (MEO), or above the highest belt (GEO) (see Figure A1.1).

Tables A1.1 and A1.2 show the ΔV required to go from (co-planar) circular LEO to circular MEO or GEO orbits by means of a Hohmann transfer orbit (see Figure A1.2). The mass fraction is that fraction of the initial LEO mass that actually gets to the desired orbit—the rest is expended propellants. The tables give semi-major axis (SMA), apogee radius (RA), perigee radius (RP), velocity at apogee (VA), velocity at perigee (VP), and ΔV to change from one orbit to

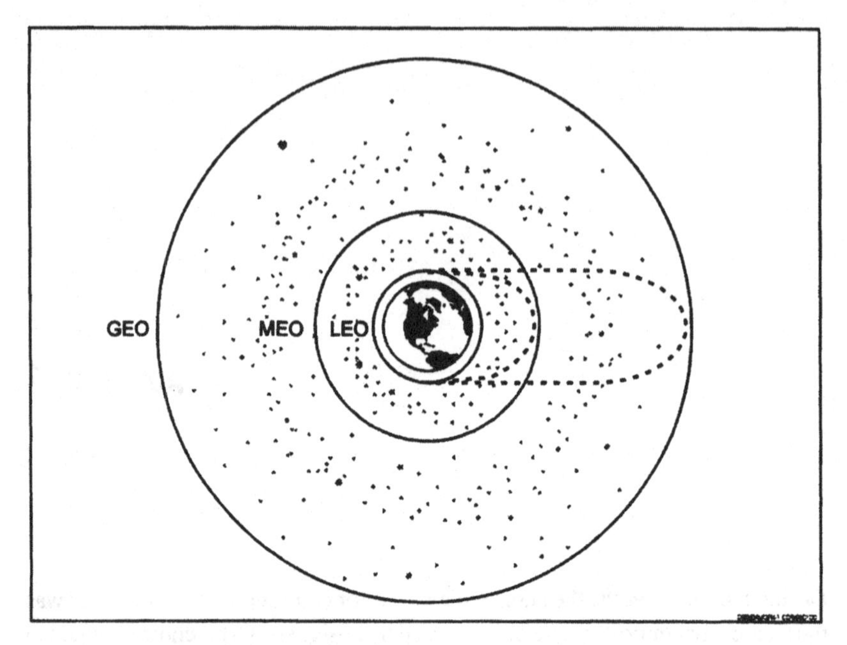

Figure A1.1. LEO, MEO, and GEO Orbits.

Table A1.1
Transfer from LEO to MEO Orbit

Orbit	SMA (km)	RA (km)	RP (km)	VA (km/s)	VP (km/s)	ΔV
Initial	6,778	6,778	6,778	7.669	7.669	
Transfer	11,578	16,378	6,778	3.775	9.121	1.452
Final	16,378	16,378	16,378	4.933	4.933	1.159
				Total ΔV		2.611
				Mass Fraction		0.411

the next. Note that an MEO satellite would be less than half the mass of an LEO satellite launched by the same rocket. A GEO satellite would be a little more than one-quarter the mass of an LEO satellite launched by the same rocket.

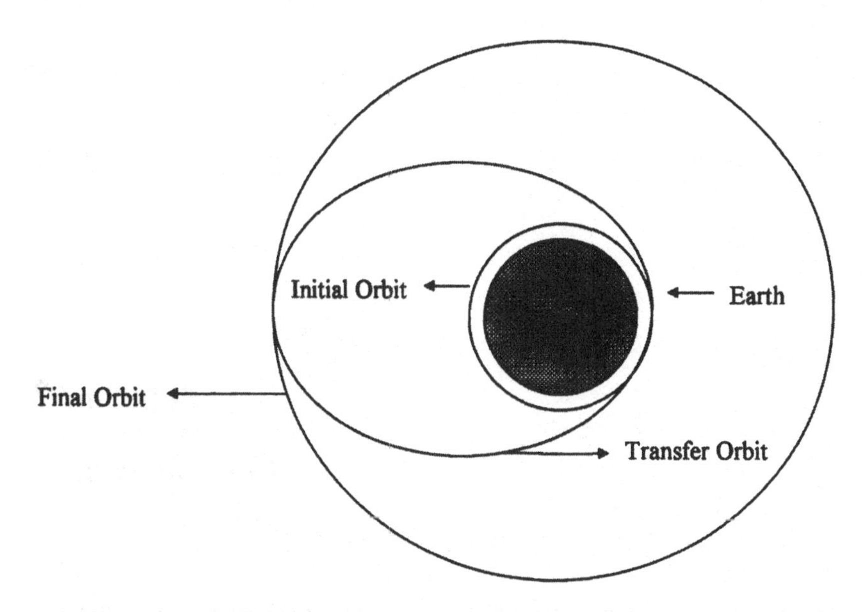

Figure A1.2. Hohmann Transfer Geometry.

Table A1.2
Transfer from LEO to GEO Orbit

Orbit	SMA (km)	RA (km)	RP (km)	VA (km/s)	VP (km/s)	ΔV
Initial	6,778	6,778	6,778	7.669	7.669	
Transfer	24,471.5	42,165	6,778	1.618	10.066	2.398
Final	42,165	42,165	42,165	3.075	3.075	1.456
					Total ΔV	3.854
					Mass Fraction	0.270

Appendix 2

Communications

The two most important elements in a communications satellite downlink are the satellite EIRP and the Earth station G/T. The important elements in the up-link are the Earth station EIRP and the satellite G/T. The downlink tends to dominate the communications problem. C/N_0 is a useful measure of performance. Subtracting 10 dB from C/N_0 gives an approximate estimate of sustainable data rate through the link.

C/N_0 = EIRP + G/T - k - PL - OL,
where
EIRP = power in dBW,
G/T = gain over noise temperature in dB/K,
k (Boltzmann's constant) = -228.3 dBW/K/Hz,
PL (path loss) = $10*\log\{(\frac{1}{4}\pi)(\lambda/R)2\}$dB [R = distance between satellite and Earth station; λ = wavelength, approximately .075m for 4 GHz C-band], and
OL (other losses) = -5dB.

Table A2.1 shows the communications data rates sustainable with a 2-watt (3 dBW) satellite transmitter, no losses, and a 6-dB gain toroidal antenna (EIRP = 9dB) at LEO, MEO, and GEO distances. The Earth station is of large size (30m) and high quality—typical of Intelsat heavy-route antennas (G/T = 40 dB/K). Modern (uncompressed) voice communications is assumed to require a 64-kbps channel; compressed digital television requires 8 Mbps. The differences in data rate are attributable strictly to the differences in range—and therefore power at

Table A2.1

Data Rate for a Given Satellite Circuit at LEO, MEO, and GEO

	Range (km)	G/T	EIRP	k	PL	OL	C/N	Data Rate	Data Rate
LEO	1,000	40	9	-228.3	-163.5	5	108.8	98.8 dBps	7.6 Gbps
MEO	15,000	40	9	-228.3	-187.1	5	85.2	75.2 dBps	33.1 Mbps
GEO	15,000	40	9	-228.3	-195.6	5	76.7	66.7 dBps	4.7 Mbps

Table A2.2

Data Rate for Different EIRP and G/T Combinations at GEO

Range (km)	G/T	EIRP	k	PL	OL	C/N	Data Rate	Data Rate
40,000	40	19	-228.3	-195.6	5	86.7	76.7 dBps	46.8 Mbps
40,000	30	29	-228.3	-195.6	5	86.7	76.7 dBps	46.8 Mbps
40,000	20	39	-228.3	-195.6	5	86.7	76.7 dBps	46.8 Mbps

the ground station antenna. Note that the "footprint" of the satellites is larger as their orbit is higher.

In later years Intelsat satellites had antennas pointing at Earth, greatly increasing EIRP, to approximately 19 dBW. These directional antennas concentrated all of the power on Earth rather than sending most of the power off into space, as had been the case with *Early Bird.* The first domestic/regional satellites had EIRPs in excess of 29 dBW. Current satellites typically have EIRPs in excess of 39 dBW. These EIRPs are paired with typical antenna G/Ts for each service: 40 dB/K (20–30m), 30 dB/K (~10m), 20 dB/K (2–3m). Note that modern VSAT systems can support the same data rates as the first Intelsat satellites with pointing antennas—and in all cases these rates are much higher than those obtainable from *Early Bird.*

Notes

Chapter 1. Introduction

1. Webb to E. Emme, September 22, 1964, NASA History Office, Washington, D.C. (hereafter cited as NHO).
2. Marcia S. Smith, "Civilian Space Applications: The Privatization Battleground," in *Space Policy Reconsidered,* ed. Radford Byerly Jr. (Boulder, Colo.: Westview Press, 1989), 105–16.
3. Peter Cunniffe, "Misreading History: Government Intervention in the Development of Commercial Communications Satellites," report #24, Program in Science and Technology for International Security, MIT, Cambridge, May 1991.
4. John R. Pierce, "Orbital Radio Relays," *Jet Propulsion,* April 1955, 44. This article was written from notes of a presentation that Pierce delivered in 1954.
5. An overview of economists' evolving views of technology and economic growth is presented in W. W. Rostow, *Theorists of Economic Growth from David Hume to the Present* (New York: Oxford University Press, 1990). See also Nathan Rosenberg, *Inside the Black Box: Technology and Economics* (Cambridge: Cambridge University Press, 1982).
6. A discussion of invention and innovation from the point of view of various historians of technology is contained in Chapter 2, "Emerging Technology and the Mystery of Creativity," in John M. Staudenmaier, *Technology's Storytellers: Reweaving the Human Fabric* (Cambridge: MIT Press, 1985).
7. F. M. Scherer, "Invention and Innovation in the Watt-Boulton Steam Engine Venture," *Technology and Culture,* spring 1965, 165–87. Reprinted in F. M. Scherer, *Innovation and Growth* (Cambridge: MIT Press, 1989), 8–31.
8. Ibid., 26.

9. "The Poor and the Rich," *The Economist*, May 25, 1996, 24.
10. A description of Hamilton and Hamiltonian thought may be found in Forrest McDonald, *Alexander Hamilton: A Biography* (New York: W. W. Norton, 1982), and Michael Lind, *Hamilton's Republic* (New York: Free Press, 1997).
11. John A. Alic, Lewis M. Branscomb, Harvey Brooks, Ashton B. Carter, and Gerald L. Epstein, *Beyond Spinoff: Military and Commercial Technologies in a Changing World* (Cambridge: Harvard University Business School Press, 1992).
12. In the 1960s, NASA contributed greatly to expenditures on electronics and aerospace procurement.
13. Both RAND and the British Interplanetary Society had been studying satellite communications since 1945, but much of their work was either unpublished or saw limited distribution.
14. GEO altitude is 36,000 kilometers above the *surface* of Earth. GEO semi-major axis is 42,000 kilometers from the *center* of Earth.
15. The Bell Labs were the final embodiment of a Bell technological drive that had begun in the 1880s, when the expiration of Bell's original patents was foreseen.
16. Arthur C. Clarke, *How the World Was One* (New York: Basic Books, 1982), was dedicated "to the real fathers of the communications satellite—John Pierce and Harold Rosen—by the Godfather."
17. 29 Fed. Ct. (1993).
18. Eugene Emme, "A Chronology of Communications Satellites" (HHR-11), April 22, 1963, NHO.
19. Edgar W. Morse, "Preliminary History of the Origins of Project Syncom," NASA Historical Note no. 44 (HHN-44), September 1, 1964, NHO.
20. George Raynor Thompson, "NASA's Role in the Development of Communications Satellite Technology," unpublished NASA Historical Manuscript no. 8 (HHM-8), [November?] 1965, NHO.
21. James E. Webb to Willis Shapley, August 11, 12, 1965, NHO.
22. Arnold W. Frutkin to Eugene M. Emme, December 30, 1965, NHO.
23. Walter A. Radius to ADA/Willis Shapley, December 10, 1965, NHO.
24. John D. Iams to Arnold Frutkin, December 9, 1965, NHO.
25. Lou Vogel to Eugene M. Emme, December 13, 1965, NHO.
26. Leonard Jaffe to Eugene M. Emme, February 24, 1966, NHO.
27. Robert G. Nunn to Eugene M. Emme, November 29, 1965, NHO.
28. J. R. Pierce to Eugene M. Emme, January 7, 1966, NHO.
29. Ronald J. Weitzel, "The Origins of ATS," NASA Historical Note no. 83 (HHN-83), August 28, 1968, NHO.
30. Jonathan F. Galloway, *The Politics and Technology of Satellite Communications* (Lexington, Mass.: Lexington Books, 1972).
31. Delbert D. Smith, *Communication via Satellite: A Vision in Retrospect* (Boston: A. W. Sijthoff, 1976).
32. Ibid., vii.
33. Ibid., viii.

34. Ibid., 261.
35. Ellis Rubinstein, "Dollars vs. Satellites," *IEEE Spectrum*, October 1976, 75–80.
36. J. R. Pierce to E. Rubinstein, November 2, 1976, AT&T Archives, Warren, N.J. (hereafter cited as AT&T).
37. K. G. McKay to Ellis Rubinstein, November 12, 1976, AT&T.
38. Ellis Rubinstein to J. R. Pierce, December 9, 1976, AT&T.
39. J. R. Pierce to K. G. McKay, November 23, December 17, 1976, AT&T; Arno A. Penzias to J. R. Pierce, July 8, 1976, AT&T; E. F. O'Neill to J. R. Pierce, August 18, 1977, AT&T; J. R. Pierce to E. Rubinstein, December 17, 1976, AT&T.
40. Pamela E. Mack, *Viewing the Earth: The Social Construction of the Landsat Satellite System* (Cambridge: MIT Press, 1990).
41. Pamela Mack, "Satellites and Politics: Weather, Communications, and Earth Resources," published later in *A Spacefaring People: Perspectives on Early Space Flight*, ed. Alex Roland (Washington, D.C.: Scientific and Technical Information Branch, NASA, 1985), 35. Mack states that Hughes won the resulting contract competitively. This is not so. RCA's Relay won the competition, as she states in the same paragraph.
42. Ibid.
43. Smith, "Civilian Space Applications," 106.
44. Linda R. Cohen and Roger G. Noll, "The Applications Technology Satellite Program," in *The Technology Pork Barrel*, ed. Linda R. Cohen and Roger G. Noll (Washington, D.C.: Brookings Institution, 1991), 149–77.
45. The authors did not seem to realize that Indonesia had the third DOMSAT system after Canada and that *Palapa* was launched in 1976. The other satellites mentioned were launched in the 1980s.
46. Cunniffe, "Misreading History," 8.
47. Ibid., 63–64.
48. José Ortega y Gasset, *La Rebelión de Las Masas* (Madrid: Espasa-Calpe, 1976), 126 (originally published in the Madrid newspaper *El Sol* in 1929–30).
49. Richard E. Neustadt and Ernest R. May, *Thinking in Time* (New York: Free Press, 1986), xi–xii.
50. Richard R. Nelson, *Government and Technical Progress* (New York: Pergamon Press, 1982), vii.

Chapter 2. From World War II to Sputnik

1. Frederick Lewis Allen, *The Big Change: America Transforms Itself, 1900–1950* (1952; First Perennial Library ed., New York: Harper and Row, 1986), 250.
2. The term "space cadets" refers to space enthusiasts, specifically those who saw the need for a progression of space missions (earth satellites followed by man in space, space station, space shuttle, Moon landing, Moon base, Mars landing). They had an "agenda" and supported immediate goals only to reach their ultimate ends.
3. Erik Barnouw, *Tube of Plenty: The Evolution of American Television*, rev ed. (Oxford: Oxford University Press, 1982), 99–100.

4. In the immediate postwar period, Western Union International and ITT controlled most of the telegraph cables, RCA controlled most of the radiotelegraphy business, and AT&T controlled most of the radiotelephony business.
5. Quoted in Wayne Biddle, *Barons of the Sky* (New York: Simon and Schuster, 1991), 288.
6. Eugene E. Bauer, *Boeing in Peace and War* (Enumclaw, Wash.: TABA, 1990), 147.
7. The events of Truman's life are admiringly recounted in David McCullough, *Truman* (New York: Simon and Schuster, 1992).
8. See Frank Kofsky, *Harry S. Truman and the War Scare of 1948* (New York: St. Martin's Press, 1995).
9. Harvey Hall, "Early History and Background on Earth Satellites," ONR:405:HH:dr, November 29, 1957, quoted in Hall, "Early U.S. Satellite Proposals."
10. Project RAND had been formed in late 1945 by the Douglas Aircraft Corporation to advise the Army Air Forces. It became an independent nonprofit R&D organization in 1948.
11. RAND, "Preliminary Design of an Experimental World-Circling Spaceship," SM-11827, May 12, 1946, RAND, Santa Monica, Calif.
12. Merton E. Davies and William R. Harris, *RAND's Role in the Evolution of Balloon and Satellite Observation Systems and Related U.S. Space Technology* (Santa Monica, Calif.: RAND, 1988), 9–19.
13. RAND, "Satellite to Surface Communication: Equatorial Orbit," RAND RM-603, July 1949, RAND, Santa Monica, Calif.
14. Paul Kecskemeti, "The Satellite Rocket Vehicle: Political and Psychological Problems," RAND RM-567, October 4, 1950.
15. Walter A. McDougall, *The Heavens and the Earth: A Political History of the Space Age* (New York: Basic Books, 1985), 108.
16. Frank H. Winter, *Prelude to the Space Age: The Rocket Societies, 1924–1940* (Washington, D.C.: Smithsonian Institution Press, 1983), 73–85.
17. Clayton R. Koppes, *JPL and the American Space Program: A History of the Jet Propulsion Laboratory* (New Haven: Yale University Press, 1982), 1–29.
18. See the popular account by Richard Rhodes, *The Making of the Atomic Bomb* (New York: Simon and Schuster, 1986), especially the epilogue. Herbert York has also covered the making of nuclear weapons from the viewpoint of an insider in *Making Weapons, Talking Peace* (New York: Basic Books, 1987). York described two of the principal actors in *The Advisors: Oppenheimer, Teller, and the Superbomb* (San Francisco: Freeman, 1976).
19. Michael H. Gorn, *Harnessing the Genie: Science and Technology Forecasting for the Air Force, 1944–1986* (Washington, D.C.: Office of Air Force History, U.S. Air Force, 1988), 11–50.
20. RAND, "A Comparison of Long Range Surface-to-Surface Rocket and Ramjet Missiles," R-174, 1950, Santa Monica, Calif., RAND.
21. Jacob Neufeld, *The Development of Ballistic Missiles in the United States Air Force, 1946–1960* (Washington, D.C.: Office of Air Force History, United States Air Force, 1990), 65–79.
22. Ibid., 98–103.

23. The story of the German rocket team is told in Frederick I Ordway III and Mitchell R. Sharpe, *The Rocket Team* (New York: Thomas Y. Crowell, 1979).
24. The use of the V-2 as a sounding rocket is detailed in David H. DeVorkin, *Science with a Vengeance* (New York: Springer-Verlag, 1992). The Viking story is told in Milton W. Rosen, *The Viking Rocket Story* (New York: Harper, 1955).
25. Wernher von Braun, *The Mars Project,* trans. Henry J. White (Urbana: University of Illinois Press, 1953).
26. Ernst Stuhlinger, "Gathering Momentum," in *Blueprint for Space: Science Fiction to Science Fact,* ed. Frederick I. Ordway III and Randy Liebermann (Washington, D.C.: Smithsonian Institution Press, 1992), 119.
27. Cornelius Ryan: *Across the Space Frontier* (New York: Viking, 1952) and *Conquest of the Moon* (New York: Viking, 1953).
28. Randy Liebermann, "The Collier's and Disney Series," in Ordway and Liebermann, *Blueprint for Space*, 135–46.
29. L. R. Shepherd, "Prelude and First Decade, 1951–1961," *Acta Astronautica* 32, nos. 7/8 (July/August 1994): 475–99.
30. Rip Bulkeley, *The Sputniks Crisis and the Early United States Space Policy* (Bloomington: Indiana University Press, 1991), 125–44; John E. Naugle, *First Among Equals* (Washington, D.C.: GPO, 1991), 6.
31. John P. Hagen, "Viking and Vanguard," in Emme, *History of Rocket Technology*, 122–27.
32. Alan S. Milward, *War, Economy, and Society, 1939–1945* (Berkeley: University of California Press, 1977), 347.
33. Microwaves are electromagnetic radiation with wavelengths between one meter and one millimeter—300 MHz to 300 GHz. Sometimes this designation is used to refer to frequencies above 1 GHz.
34. Philip L. Cantelon, "The Origins of Microwave Telephony," *Technology and Culture*, July 1995, 560–82.
35. U.S. Department of Commerce, *1984 World's Submarine Telephone Cable Systems* (Washington, D.C.: GPO, 1984), 90–91.
36. Ibid., 101.
37. Ibid., 127.
38. Hugh G. J. Aitken, *The Continuous Wave: Technology and American Radio, 1900–1932* (Princeton: Princeton University Press, 1985), 481–82.
39. Arthur C. Clarke, "Extra-Terrestrial Relays," *Wireless World* 51, no. 10 (October 1945): 303–8.
40. Arthur C. Clarke, *The Exploration of Space* (New York: Harper and Row, 1952). Many of the articles and books mentioned in this work have two dates associated with them. The first date usually reflects when they were first available and the second when they were published. In this case, Clarke's book was first available in the United Kingdom in 1951 and then published in 1952 in the United States.
41. Homer E. Newell, *Beyond the Atmosphere: Early Years of Space Science*, NASA SP-4211 (Washington, D.C.: GPO, 1980), 106.

42. See note #16, chapter 1.
43. Eric Burgess, "The Establishment and Use of Artificial Satellites," *Aeronautics*, September 1949, 70–82.
44. Scientists and engineers coming from the world of aircraft and high-altitude experiments measure orbits from the *surface* of Earth. Orbital analysts and astronomers measure from the *center* of Earth. Orbits less than 6,378 kilometers (Earth's radius) can be assumed to be measured from the surface. All other orbital "heights" can be assumed to be measured from the center of Earth, unless referred to as *altitudes*.
45. John R. Pierce, "Orbital Radio Relays," *Jet Propulsion,* April 1955.
46. R. P. Haviland, "The Communication Satellite," in *Eighth International Astronautical Congress Proceedings* (Vienna: Springer-Verlag, 1958), 543–62.
47. *New York Times,* October 6, 1957.
48. McDougall, *The Heavens and the Earth*, 140.
49. *Public Papers of the Presidents, Dwight D. Eisenhower 1953*, 182, quoted in ibid., 114.

Chapter 3. Post-Sputnik

1. Dated events, unless specifically footnoted, are from U.S. Congress, House, *A Chronology of Missile and Astronautic Events [1915–1960],* 87th Cong., 1st sess., 1961.
2. U.S. Congress, House, Select Committee on Astronautics and Space Exploration, *The National Space Program,* 85th Cong., 2d sess., 1958 34.
3. Biographical data on Glennan is from Roger A. Launius, introduction to *The Birth of NASA: The Diary of T. Keith Glennan,* ed. J. D. Hunley (Washington, D.C.: NASA History Office, 1993), x–xi.
4. Ibid., x–xviii.
5. Donald C. Elder, *Out from behind the Eight-Ball: A History of Project Echo* (San Diego: American Astronautical Society, 1995), p.25.
6. John R. Pierce and Rudolf Kompfner, "Transoceanic Communications by Means of Satellites," *Proceedings of the IRE* , March 1959, 372–80.
7. Pierce, *Beginnings*, 9–12.
8. Ibid., 12–13; J. R. Pierce to C. C. Cutler, October 17, 1958, AT&T.
9. C. C. Cutler to J. R. Pierce, October 27, 1958, AT&T.
10. T. Keith Glennan to James H. Douglas, August 18, 1960, NHO. This letter refers to the earlier agreement and requests a change to allow NASA to proceed with an active satellite.
11. U.S. Congress, House, Select Committee on Astronautics and Space Exploration, *The Next Ten Years in Space, 1959–1969,* 86th Cong., 1st sess., 1959.
12. Ibid., 211.
13. Ibid., 118–24.
14. Ibid., 221.
15. Ibid., 32.
16. Ibid., 35, emphasis added.

17. John Naugle, memorandum to Administrator, "Recommendation for Executive Performance Award," July 28, 1975, NHO.
18. Roy W. Johnson to T. Keith Glennan, December 17, 1958, NHO.
19. T. Keith Glennan to Roy W. Johnson, January 5, 1959, NHO.
20. Both the JPL and AT&T antennas had distinguished careers after Echo. The AT&T antenna was used to discover the cosmic background radiation, and the JPL Goldstone antenna became part of the Deep Space Network used to track planetary probes.
21. J. R. Pierce, R. Kompfner, and C. C. Cutler, "Memorandum for the Record: Research toward Satellite Communication, Case 38543," January 6, 9, 1959, AT&T; J. R. Pierce to J. W. McRae, January 7, 1959, AT&T.
22. Harold A. Rosen, "Harold Rosen on Satellite Technology Then and Now," *Via Satellite*, July 1993, 40–43; GM–Hughes Electronics, "History and Accomplishments of the Hughes Aircraft Company," n.d., Hughes Aircraft Company Archives, El Segundo, Calif. (hereafter cited as HAC), 12; Edgar W. Morse, "Preliminary History of the Origins of Project SYNCOM," NASA Historical Note no. 44 (HHN-14), September 1, 1964, NHO, 32–34,.
23. U.S. Congress, House, Committee on Science and Astronautics, *Hearings: Satellites for World Communication,* 86th Cong., 1st sess., 1959.
24. Ibid., 2–34; U.S. Congress, House, Committee on Science and Astronautics, *Hearings: Project Advent-Military Communications Satellite Program,* 87th Cong., 2d sess., 1962, 90–91.
25. House, Committee on Science and Astronautics, *Hearings: Satellites for World Communication*, 35–40.
26. Ibid., 40–46.
27. Ibid., 47–56.
28. Ibid., 56–62.
29. Ibid., 97–106.
30. U.S. Congress, House, Committee on Science and Astronautics, *Report: Satellites for World Communication,* 86th Cong., 1st sess., 1959.
31. Pierce, J. R. (?), "BTL Tracking Proposal," April 2, 1959, AT&T; W. C. Jakes, "Visit to Washington on March 31, 1959," April 7, 1959, AT&T; J. R. Pierce to J. A. Morton, April 14, 1959, AT&T.
32. Philip J. Klass, "Civil Communication Satellites Studied," *Aviation Week*, June 22, 1959, 189–97.
33. "U.S. Will Develop Radio Satellites," *New York Times*, August 29, 1959, 14.
34. A. S. Jerrems to F. R. Carver, September 17, 1959, HAC.
35. S. G. Lutz to A. V. Haeff, "Evaluation of H. A. Rosen's Commercial Satellite Communication Proposal," October 1, 1959, HAC.
36. Ibid., 6.
37. S. G. Lutz to A. V. Haeff, "Economic Aspects of Satellite Communication," October 13, 1959, HAC.
38. The task force working members were E. D. Felkel, S. G. Lutz, D. E. Miller, H. A. Rosen, and J. H. Striebel.

39. S. G. Lutz to A. V. Haeff, "Commercial Satellite Communication Project: Preliminary Report of Study Task Force," October 22, 1959, HAC.
40. J. H. Striebel to A. V. Haeff, "Market Study of a World Wide Communication System for Commercial Use," October 22, 1959, HAC.
41. L. A. Hyland to A. E. Haeff, C. G. Murphy, "Communications Satellite," October 26, 1959, HAC.
42. S. G. Lutz to A.V. Haeff, "Satellite Data," October 29, 1959, HAC.
43. H. A. Rosen and D. D. Williams, "Commercial Communication Satellite," October 1959, Airborne Systems Laboratories, Hughes Aircraft Company, HAC.
44. David F. Doody to L. A. Hyland, "Commercial Communication Satellite (H. A. Rosen and D. D. Williams) Patentable Novelty," October 29, 1959, HAC.
45. The last act in this dispute occurred in 1993 when the U.S. Court of Claims ruled that the U.S. government owed Hughes royalties of as much as $1 billion. See Edmund L. Andrews, "Big Award Is Ordered for Hughes," *New York Times*, August 18, 1995, D5.
46. D. D. Williams to D. F. Doody, "Discussions with Dr. Homer J. Stewart, NASA," November 23, 1959, HAC.
47. In his autobiography *Call Me Pat* (Virginia Beach, Va.: Donnington, 1993), L. A. Hyland describes how Williams brought him a certified check for $10,000 as Williams's personal contribution to the development of the synchronous satellite. Although Hyland returned the check, it made him wonder if he had become too risk-averse.
48. David F. Doody to Noel Hammond, "Communication Satellite," December 1, 1959, HAC.
49. H. A. Rosen and D. D. Williams, "Commercial Communication Satellite," RDL/B-1, January 1960, Engineering Division, Hughes Aircraft Company, HAC.
50. J. W. Ludwig and J. H. Striebel to R. K. Roney, "Commercial Communications Satellite—Logistics," December 22, 1959, HAC.
51. AT&T/BTL, "Project Echo: Monthly Report No. 3, Contract NASW-110, December 1959," AT&T.
52. Pierce, *Beginnings*, 15–22; A. C. Dickieson, "TELSTAR: The Management Story," unpublished monograph, Bell Telephone Laboratories, July 1970, 29–32, AT&T.
53. Pierce, *Beginnings*, 22–23; Dickieson, "TELSTAR," 32–34.
54. John B. Medaris, *Countdown for Decision* (New York: Putnam, 1960), 249–95. Medaris later became an Episcopalian priest.
55. Dinter H. Schwebs (IDA/ARPA), "Memorandum for the Record: Reorientation of NOTUS Program," December 8, 1959, NHO.
56. John W. Finney, "Speed-Up Urged on Three Satellites," *New York Times*, March 31, 1960, 9.
57. Robert K. Roney to A. E. Puckett, "Communications Satellite Review Analysis," January 27, 1960, HAC; R. L. Corbo to T. W. Oswald, "Revision of the Communication Satellite Structure and General Arrangement," January 29, 1960, HAC; Appendix B to Rosen and Williams, "Commercial Communications Satellite," October 1959.
58. H. A. Rosen and Tom Hudspeth, interview, California Museum of Science and Technology, spring 1992, HAC; Allen E. Puckett to D. F. Doody, "Release of

Hughes' Rights in Inventions Disclosed Relating to Communications Satellite," March 7, 1960, HAC; Allen E. Puckett to L. A. Hyland, "Communication Satellite," March 21, 1960, HAC.

59. Dickieson, "TELSTAR," 34–39.
60. Allen E. Puckett to Richard S. Morse, April 8, 1960, HAC; J. W. Ludwig to C. G. Murphy/E. A. Puckett [*sic*], "Communication Satellite—Presentation to Dr. E. G. Wittimg, Deputy Director, R and D, U.S. Army," May 2, 1960, HAC; J. W. Ludwig to A. E. Puckett, "Army RFP on Advent System, Assignment of HAC Responsibility," May 19, 1960, HAC; F. D. Vieth to Lt. Col. L. B. Brownfield, May 27. 1960, HAC.
61. U.S. Congress, Senate, NASA Authorization Subcommittee of the Committee on Aeronautical and Space Sciences, *Hearings on NASA Authorization for Fiscal Year 1961,* 86th Cong., 2d sess., 1960, 69–70.
62. U.S. Congress, Senate, Committee on Aeronautical and Space Sciences, *Staff Report on Communications Satellites: Technical, Economic, and International Developments,* 87th Cong., 2d sess., 1962, 192; Dickieson, "TELSTAR," 2–7.
63. S. G. Lutz to File, "AFCEA Papers by B.T.L. on Satellite Communication," June 3, 1960, HAC.
64. S. G. Lutz to File, "Conference with Leon Jaffee [*sic*], NASA, Re Satellite Communication, Frequency Sharing," May 31, 1960, HAC.
65. "Comsat Financial Status: 3 July 1960," July 13, 1960, HAC; "Commercial Communication Satellite, Preliminary Cost Estimates," September 14, 1960, HAC.
66. Hughes Aircraft Company, "Synchronous Communication Satellite, Proposed NASA Experimental Program, Exhibit B: Technical Discussion," ED 655R, 60H-6702/5874-005, June 1960, HAC; Hughes Aircraft Company, "Comparison of the Hughes Synchronous Repeater with the BTL Low Altitude Repeater," n.d., HAC; Hughes Aircraft Company, "Advantages of Stationary Satellite Systems," n.d., HAC.
67. In the confusing world of mergers and divestitures, Ford Aeronutronic may not be recognizable as the Ford part of the mid-1960s Philco-Ford Company. The Ford section later became Ford Aerospace, then Loral, and is now part of Lockheed-Martin. Philco—now Space Systems Loral—built the "first" communications satellite and became one of the "big three" communications satellite manufacturers.
68. D. D. Williams, "Dynamic Analysis and Design of the Synchronous Communication Satellite," TM-649, May 1960, HAC; D. D. Williams and P. Wong to F. R. Carver, "Orbit Determination for Satellite Surveillance System," August 18, 1960, HAC.
69. Allen E. Puckett to Those Listed, "Communication Satellite Presentation," July 26, 1960, HAC; John S. Richardson to C. G. Murphy, "Communications Satellite," August 12, 1960, HAC; F. P. Adler to J. W. Ludwig, H. A. Rosen, "Visit of Mr. Jaffe," August 26, 1960, HAC; "Conference on Communication Satellite Program at General Telephone Offices Menlo Park," September 15, 1960, HAC; C. G. Murphy to H. A. Rosen, "Visit of Dr. S. Reiger and Other RAND Corporation Personnel," September 21, 1960, HAC; A. E. Puckett to [J. W.] Ludwig, [H. A.] Rosen, "Visit from ITT Laboratories Regarding Communication Satellite," November 18, 1960, HAC; "Chronology of Communication Satellite Program," n.d., HAC.

70. NASA, *Fourth Semi-Annual Report to Congress* (Washington: GPO, 1961), 10–17.
71. G. L. Best, "Memorandum: Satellite Communications," September 6, 1960, AT&T.
72. G. L. Best to T. Keith Glennan, September 15, 1960, AT&T; T. Keith Glennan to G. L. Best, September 28, 1960, AT&T; E. I. Green to T. Keith Glennan, October 20, 1960, AT&T; Dickieson, "TELSTAR," 8, 40–44.
73. A. W. Betts to J. H. Rubel, "Communications Satellite," June 13, 1960, personal files of John H. Rubel, Tesuque, N.M. (hereafter cited as JHR); Richard Witkin, "Signal Satellites Due for Decision," *New York Times*, June 23, 1960, 10; John H. Rubel to A. S. Jerrems, September 22, 1960, HAC; "Space Projects Shifted," *New York Times*, September 23, 1960, 2; John H. Rubel to T. Keith Glennan, September 27, 1960, JHR; H. A. Rosen to E. G. Witting, October 25, 1960, HAC; Edward E. Harriman, "Project Summary on Courier," November 18, 1960, NHO.
74. *Diary of T. Keith Glennan,* 142.
75. Ibid., 189.
76. Glennan activities are from ibid., 189–207.
77. Later the Office of Management and Budget (OMB).
78. *Diary of T. Keith Glennan*, 207–10.
79. Robert C. Seamans Jr., *Aiming at Targets* (Washington, D.C.: NASA, 1996).
80. NASA, press release, September 1, 1960, NHO.
81. *Diary of T. Keith Glennan*, 232–34.
82. Ibid., 237, 244–45, 253.
83. Ibid., 249, 256–57.
84. Memorandum (no name, no date, presumably Nunn's), NHO; R. E. Warren, memorandum, October 14, 1960, NHO; Robert G. Nunn, memorandum, October 28, 1960, NHO.
85. William Meckling and Siegfried Reiger, "Communications Satellites: An Introductory Survey of Technology and Economic Promise," RAND Corporation, Report RM-2709-NASA, September 15, 1960, RAND, Santa Monica, Calif.
86. *Diary of T. Keith Glennan*, 286.
87. Ibid., 278–92.
88. Robert G. Nunn, "Memorandum for the Record," December 23, 1960, NHO.

Chapter 4. Government Intervenes

1. The general consensus seems to be that Kennedy agreed with Eisenhower's view that space was probably a waste of money except for specific, primarily military, applications. Eisenhower's opinion is discussed in Charles Murray and Catherine Bly Cox, *Apollo: The Race to the Moon* (New York: Simon and Schuster, 1989), 61.
2. U.S. Congress, House, Committee on Science and Astronautics, *Defense Space Interests,* 87th Cong., 1st sess., 1961, 22–23.
3. George B. Kistiakowsky, *A Scientist at the White House* (Cambridge: Harvard University Press, 1976), 52.
4. Apparently ITT had made a similar offer. See *The Birth of NASA: The Diary of T.*

Keith Glennan, ed. J. D. Hunley. (Washington, D.C.: NASA History Office, 1993), 301.

5. Ibid., 303.
6. Robert C. Seamans Jr., NASA Exit Interview, May 8, May 30, and June 3, 1968, NHO, 17–19.
7. Webb had participated in the 1947 DoD budget slashing, which canceled, among other programs, the Atlas ICBM.
8. Walter A. McDougall, *The Heavens and the Earth: A Political History of the Space Age* (New York: Basic Books, 1985), 311.
9. NASA, *Aeronautical and Astronautical Events of 1961* (Washington, D.C.: GPO, 1962), 4–8; Minutes: Administrator's Staff Meeting, January 18, 26, February 2, 1961, NHO.
10. Abe Silverstein to Assistant Directors et al., "Fiscal Year 1963 Preliminary Budget Estimates: Additional Information Concerning," March 1, 1961, NHO; A. Silverstein to H. Goett, "Designations for Missions and Payloads," March 1, 1961, NHO.
11. Seamans, NASA Exit Interview, 13, 25–26; H. L. Dryden, press conference transcript, January 14, 1961, NHO, 21; Robert L. Rosholt, *An Administrative History of NASA, 1958–1963* (Washington, D.C.: GPO, 1966), 195; James E. Webb to "The Director," Bureau of the Budget, March 13, 1961, NHO.
12. James E. Webb, "Administrator's Presentation to the President," March 21, 1961, NHO.
13. Fred R. Kappel to James E. Webb, April 5, 1961, copy in *Exploring the Unknown: Selected Documents in the History of the U.S. Civil Space Program,* edited by John M. Logsdon et al., vol. 3., *Using Space* (Washington, D.C.: NASA, 1998), 45–57.
14. James E. Webb to Fred R. Kappel, April 8, 1961, copy in ibid., 58–60.
15. *Diary of T. Keith Glennan*, 290.
16. This suggestion even came from within NASA: Don Ostrander to Dr. Seamans, "Reflections on the American Posture in Space," April 21, 1961, NHO.
17. James E. Webb to John A. Johnson, April 28, 1961, NHO.
18. James E. Webb to Overton Brooks, May 2, 1961, NHO.
19. Robert G. Nunn, "Memorandum for Record," May 5, 1961, NHO.
20. Fred R. Kappel to James E. Webb, May 8, 1961, copy in Logsdon et al., *Exploring the Unknown.*
21. Director of Defense Research and Engineering (DDRE), "Report: Management of Advent," April 3, 1960, JHR.
22. James H. Douglas (acting Secretary of Defense), to Secretary of the Army and Secretary of the Air Force, "Program Management for ADVENT," September 15, 1960, JHR.
23. U.S. Congress, House, Committee on Science and Astronautics, *Hearings: Communications Satellites, Part I,* 87th Cong., 1st sess., 1961, 1.
24. Ibid., 60.
25. Ibid., 102.
26. Ibid., 306.
27. Robert C. Seamans Jr. to [J. E.] Webb and [H. L.] Dryden, "Status of Planning for an Accelerated NASA Program," May 12, 1961, NHO.

28. Robert C. Seamans Jr., *Aiming at Targets* (Washington, D.C.: NASA, 1996), 88–90.
29. Robert G. Nunn, "Memorandum for the Associate Administrator," May 16, 1961, NHO.
30. NASA, news release, "Statement by James E. Webb, Administrator," #61-112, May 25, 1961; NASA, news release, "Budget Briefing," #61-115, May 25, 1961; Minutes: Administrator's Staff Meeting, May 25, 1961, NHO.
31. Minutes: Administrator's Staff Meeting, June 1, 1961, NHO.
32. James E. Webb to Robert S. McNamara, June 1, 1961, NHO.
33. Minutes: Administrator's Staff Meeting, June 8, 1961, NHO.
34. James E. Webb, "Memorandum for Dr. Dryden," June 12, 1961, NHO.
35. Minutes: Administrator's Staff Meeting, June 15, 1961, NHO.
36. Ibid., June 22, 1961.
37. Ibid., June 29, 1961.
38. I. Welber, "Memorandum for File," July 10, 1961, AT&T.
39. Edward C. Welsh "Memorandum for the Vice President," April 28, 1961, National Archives II, College Park, Md. (hereafter cited as NAII).
40. Andrei Gromyko to Central Committee CPSU, May 20, 1961, quoted in Aleksandr Fursenko and Timothy Naftali, *One Hell of a Gamble: Khrushchev, Castro, and Kennedy, 1958–1964* (New York: Norton, 1997), 121.
41. Edward C. Welsh "Memorandum for the Vice President," June 5, 1961, NAII.
42. John F. Kennedy, draft of letter to LBJ, n.d. [early June 1961], Welsh files, NAII.
43. Richard Hirsh, "Memorandum for Dr. Welsh: Highlights of Meeting of June 27, 1961, Concerning Communications Satellites," June 28, 1961, NAII.
44. Richard Hirsh, "Memorandum for Dr. Welsh: Highlights of Meeting of June 28, 1961, Concerning Communications Satellites," June 29, 1961, NAII.
45. Ibid.
46. Hal Taylor, "Council Favors Private Ownership," *Missiles and Rockets*, July 3, 1961, 11, 40.
47. The White House, press release, "Statement of the President on Communications Satellite Policy," July 24, 1961, NAII.
48. Humphrey, Kefauver, Morse, et al. to John F. Kennedy, August 24, 1961, reprinted in U.S. Congress, Senate, Committee on Foreign Relations, *Hearings: Communications Satellite Act of 1962,* 87th Cong., 2d sess., 1962, 51–54.
49. House, Committee on Science and Astronautics, *Hearings: Communications Satellites, Part 1*, 461.
50. Ibid., 464.
51. Ibid., 546.
52. Seamans, NASA Exit Interview, 28.
53. U.S. Congress, House, Committee on Science and Astronautics, *Hearings: Communications Satellites, Part 2,* 87th Cong., 1st sess., 1961, 739.
54. Lyndon B. Johnson to Gerald W. Groeppor, July 20, 1961, NAII.
55. Much of the information on the NASC comes from the document "National Aeronautics and Space Council" by John Mark Mobius of MIT. His references are coded,

but the code is lost, preventing proper acknowledgment of sources. This document is in the NHO files.

56. Russell W. Hale, "Memorandum for Dr. Welsh," September 23, 1961, NAII.
57. U.S. Congress, House, Committee on Science and Astronautics, *Report: Commercial Applications of Space Communications Systems,* 87th Cong., 1st sess. 1961, 1.
58. Welsh and John Johnson, the NASA General Counsel, probably contributed to writing all three bills.
59. Philip J. Farley, "Memorandum for Dr. Edward C. Welsh," November 28, 1961, NAII.
60. U.S. Congress, House, Committee on Science and Astronautics, *Hearings: Satellites for World Communications,* 86th Cong., 1st sess., 1959; U.S. Congress, House, Committee on Science and Astronautics, *Report: Satellites for World Communications,* 86th Cong., 1st sess., 1959.
61. Delbert D. Smith, *Communication via Satellite: A Vision in Retrospect* (Boston: A. W. Sijthoff, 1976), 93–103.
62. Ibid., 104.
63. U.S. Congress, House, Committee on Interstate and Foreign Commerce, *Hearings: Communications Satellites, Part 2,* 87th Cong., 1st sess., 1961, 13–21.
64. U.S. Congress, Senate, Committee on Aeronautical and Space Sciences, *Communications Satellites: Technical, Economic, and International Developments,* 87th Cong., 1st sess., 1962.
65. A major benefit to observers and analysts of the U.S. political scene is the propensity of congressional hearings and reports to include the texts of most other documents pertinent to the issue at hand.
66. Senate, Committee on Aeronautical and Space Sciences, *Communications Satellites,* 1.
67. Quoted in ibid., 30, originally in the *New York Times*, February 13, 1961, 13.
68. Eventually RCA would be the fourth-largest investor in Comsat, after AT&T, ITT, and GTE.
69. "House Votes 354-to-9 Approval of Space Communications Firm," *Washington Post,* May 4, 1962, A1.
70. U.S. Congress, House, Committee on Aeronautical and Space Sciences, *Report: Communications Satellite Act of 1962,* 87th Cong., 2d sess., 1962.
71. The political actions leading to the passage of the Communications Satellite Act of 1962 have been studied in great detail in books, articles, and dissertations. Much of the debate is detailed in the various published hearings. The description here is compiled from various sources, especially Smith, *Communication via Satellite*; Jonathan F. Galloway, *The Politics and Technology of Satellite Communications* (Lexington, Mass.: Lexington Books, 1972); and Michael E. Kinsley, *Outer Space and Inner Sanctums* (New York: John Wiley and Sons, 1976).

Chapter 5. Building the Satellites

1. Hughes Aircraft Company, "SYNCOM" (brochure), 3/62/2M, NHO.

2. [C. G. Murphy?], "Policy Statement for Exploitation of HAC Communications Satellites," [early 1962?], HAC.
3. A. C. Dickieson, "TELSTAR: The Management Story," unpublished monograph, Bell Telephone Laboratories, July 1970, AT&T, 85–130.
4. Ibid., 2.
5. Ibid., 5, 18, 19.
6. Ibid., 6–13.
7. Ibid., 85–127.
8. Ibid., 129–55.
9. Ibid., 157–91.
10. Ibid., 193–97.
11. Ibid., 199–210.
12. Ibid., 255–68.
13. "NASA's ComSat Funding to Climb," *Missiles and Rockets*, April 2, 1962, p.17.
14. Samuel G. Lutz, "Satellite Communications and Frequency Sharing," *Interavia*, June 1962, 753–57.
15. NASA, news release, "NASA Plans Follow-on Communications Satellite," #62-139, June 18, 1962, NHO.
16. Harold Brown (John Rubel) to George P. Miller, July 10, 1962, quoted in U.S. Congress, House, Committee on Science and Astronautics, *Hearings: Project Advent—Military Communications Satellite Program,* 87th Cong., 2d sess., 1962, 96–100.
17. Ibid., 90–105 (Rubel testimony).
18. Ibid., 129.
19. The Andover station was also referred to as the Rumford, Maine, station and as "Space Hill."
20. Dickieson, "TELSTAR," 154–60.
21. Ibid., 161–62.
22. Ibid., 162–76.
23. AT&T/BTL, *Project Telstar: Preliminary Report, Telstar 1*, July-September 1962, 30–41, NHO.
24. Ibid., 43–79.
25. Ibid., 85–113.
26. U.S. Congress, House, Committee on Aeronautical and Space Sciences, *Hearings: Commercial Communications Satellites*, September 18, 19, 21, 27, October 4, 1962, 87th Cong., 2d sess., 1962, 175–77.
27. Ibid., 1.
28. Ibid., 1–4.
29. Ibid., 4–62.
30. Ibid., 63–109.
31. Ibid., 111–34.
32. Ibid., 135–82.
33. Jack Raymond, "President Names Satellite Board," *New York Times*, October 5, 1962, 1.

34. Frank C. Porter, "Satellite Company Incorporators Get Briefing from U.S.," *Washington Post*, October 23, 1962.
35. "FCC Would Allow All Phone Firms to Invest in Space Message Firm," *Wall Street Journal*, November 29, 1962.
36. Drew Pearson, "The Revolution in Communications," *Washington Post*, November 26, 1962, B23; Drew Pearson, "Satellite Corp. Meets in Secrecy," *Washington Post*, November 27, 1962.
37. Robert S. Allen and Paul Scott, "Satellite Corp. Growth Pains," *Newport News, Va., Times Herald*, January 4, 1963, 10.
38. "Development of the Relay Communications Satellite, *Interavia*, June 1962, 758–59; John P. MacKenzie, Sidney Metzger, and Robert H. Pickard, "Relay," *Aeronautics and Aerospace Engineering*, September 1963, 64–67; NASA, *Final Report on the Relay 1 Program* (Washington, D.C.: GPO, 1965), 63–90.
39. NASA, news release, "Technical Background Briefing: Project Syncom," January 29, 1962, NHO.
40. "South African Observatory Spots Missing Syncom," *Washington Post*, March 1, 1963; "Observatory Pinpoints Lost Satellite Definitely," *Washington, D.C., Sunday Star*, March 3, 1963; R. Warren, "Syncom I Progress Report No. 4," March 4, 1963, NHO.
41. "NASA Sees Telstar-Type Satellite as Best for World-Wide System," *Aviation Week & Space Technology*, September 24, 1962, 40.

Chapter 6. Choosing a System

1. "Satellite Communication," *International Science and Technology*, January 1963, 69–74.
2. "Satellite Job Is Resigned by Graham," *Washington Evening Star*, January 26, 1963, A9.
3. "Satellite Firm Selects Top Officers, Hints at Stock Sale in a Year," *Wall Street Journal*, February 28, 1963; "Welch, Charyk Picked to Head Satellite Firm," *Washington Post*, March 1, 1963, A3.
4. John W. Finney, "Space Radio Plan Creating Doubts," *New York Times*, April 24, 1963, 2.
5. Cecil Brownlow, "U.S.-Europe Comsat Agreement Predicted," *Aviation Week & Space Technology*, April 22, 1963, 74–75.
6. S. H. Reiger, R. T. Nichols, L. B. Early, and E. Dews, "Communications Satellites: Technology, Economics, and System Choices," RM-3487-RC, February 1963, RAND, Santa Monica, Calif.
7. NASA, "Program Review: Communications, Meteorology, Future Applications," June 29, 1963, NHO.
8. Ibid., 3.
9. Ibid., 3–7.
10. One of the systems that failed on *Relay 1* was the automatic timer, which was

supposed to disable the satellite after one year. The satellite eventually ceased functioning in February 1965.

11. NASA, "Program Review," 8–25.
12. Ibid., 25.
13. Ibid., 35–42
14. Ibid., 42–51.
15. Ibid., 54–57.
16. Ibid., 61–64.
17. Ibid., 175–92.
18. H. L. Dryden to J. V. Charyk, June 24, 1963, NHO.
19. Robert C. Toth, "F.C.C. Sees Delay in Space Stock," *New York Times*, July 27, 1963, 3; "Satellite Corp., Prodded by FCC, May Tell More This Week about Plans for Stock Sale," *Wall Street Journal*, July 29, 1963.
20. Siegfried H. Reiger, "Commercial Satellite Systems," *Astronautics and Aerospace Engineering*, September 1963, 26–30; Joseph V. Charyk, "Communications Satellite Corporation: Objectives and Problems," *Astronautics and Aerospace Engineering*, September 1963, 45–47.
21. Paul E. Norsell, "SYNCOM," *Astronautics and Aerospace Engineering*, September 1963, 76–78.
22. NASA, news release, "Technical Background Briefing: Project Syncom," January 29, 1963, NHO.
23. R. Darcey to H. Goett, June 14, 1963, NHO.
24. Ibid., interviews with H. Rosen and G. Murphy, 50.
25. R. W. Cole to D. D. Williams, "Summary of Orbital Data for Syncom 2 (A-26), August 18, 1963, HAC.
26. L. M. Field to L. A. Hyland, "Possible Use of Syncom as a Navigational System–Microwave Loran," August 9, 1963, HAC.
27. C. G. Murphy to L. A. Hyland, "Synchronous Altitude Communication Satellite System," September 16, 1963, HAC.
28. L. A. Hyland to Robert Gilruth, October 21, 1963, HAC.
29. R. E. Warren to R. Garbarini, September 10, 1963, NHO; R. Garbarini to H. Goett, September 24, 1963, NHO.
30. U.S. Congress, House, Committee on Government Operations, Military Operations Subcommittee, *Report: Satellite Communications—Military-Civil Roles and Relationships,* 88th Cong., 2d sess., 1964, 11–12.
31. Ibid., 51–53.
32. "COMSAT System Selection Pending," *Space Daily*, March 25, 1964, 463; "Navy Prefers MA/GG COMSAT; COMSAT Act Repeal Urged by Senators," *Space Daily*, March 26, 1964, 471–72; "Two Senators Hit COMSAT, U.S. Deal," *Washington Post*, March 26, 1964, 7; "Incentive Potential of $4.8 Million in Hughes-COMSAT Corp. Contract," *Aviation Week & Space Technology*, March 30, 1964; "COMSAT Static," *Aviation Week & Space Technology*, March 30, 1964, 15; "The COMSAT Problem," *Aviation Week & Space Technology*, April 6, 1964, 11;

"House Group Studies NASA Role as COMSAT Corp. Technical Adviser," *Aviation Week & Space Technology*, April 13, 1964, 32.

33. Comsat, press release, "Note to Correspondents," December 10, 1963; James E. Dingman to Leo Welch, December 6, 1963, NHO.
34. Comsat, press release, "Commercial Communications Satellite Engineering Design Proposals Requested of Industry; Corporation Regards Move as Major Step," December 22, 1963.
35. "NASA's ComSat Funding to Climb," *Missiles and Rockets*, April 2, 1962, 17.
36. "NASA Sees TELSTAR-Type Satellite as Best for World-Wide System," *Aviation Week and Space Technology*, September 24, 1962, 40.
37. John J. Kelleher, memorandum to Leonard Jaffe, "FCC Action on Early Bird," January 24, 1964, NHO.
38. Comsat, "Report Pursuant to Section 404 (b) of the Communications Satellite Act of 1962 for the Period February 1, 1963, to December 31, 1963," January 31, 1964.
39. There were several problems associated with these teams. Philco and TRW had teamed on MACS. ITT was an adviser to DoD on MACS. It was quite possible that MACS would be canceled and COMSAT would provide the capability required by DoD.
40. "Communications Satellite Corp. Gets Bids for Designing System from Six Companies," *Wall Street Journal*, February 12, 1964, 1; Cecil Brownlow, "International Comsat Agency Considered," *Aviation Week & Space Technology*, February 17, 1964, 34.
41. "Ocean Satellite by 1965 Is Sought," *New York Times*, March 5, 1964, 4; "Permission Sought to Orbit Communications Device in '65," *Baltimore Sun*, March 5, 1964; "COMSAT Files with FCC," *Missiles and Rockets*, March 9, 1964; "Communications Satellite Firm Negotiates Spacecraft Contract with Hughes Aircraft," *Wall Street Journal*, March 17, 1964; "FCC Gives ComSat Go-Ahead," *Missiles and Rockets*, April 20, 1964, 10.
42. NASA, "Associate Administrator's Advanced Study Review: Earth Orbital Studies," June 25, 1964, A19–A20, NHO.
43. John P. MacKenzie, "Two Hundred Carriers File for COMSAT Shares," *Washington Post*, March 24, 1964, 4; "Satellite Corp. to Market Stock at $20 a Share," *Wall Street Journal*, May 7, 1964; Philip J. Klass, "COMSAT Firm Files $200-Million Fund Plan," *Aviation Week & Space Technology*, May 11, 1964, 25–26; Robert Hotz, "Selling Shares in Space," *Aviation Week & Space Technology*, May 11, 1964, 17; John P. MacKenzie, "Industry Snaps Up Its Half of COMSAT Stock," *Washington Post*, March 28, 1964, 2; "Satellite Corp. Picks Six Candidates for New Board," *New York Times*, July 9, 1964, 43C; "COMSAT Stock Widely Distributed," *Space Daily*, August 14, 1964, 227; S. Oliver Goodman, "130,000 Owners Listed in Initial Report of COMSAT," *Washington Post*, August 14, 1964, D7; John W. Finney, "COMSAT Takes On Private Role as Holders Stage First Meeting," *New York Times*, September 18, 1964, 45; Suzanne Montgomery, "COMSAT Board Awaits Senate Nod," *Missiles and Rockets*, September 28, 1964, 34.

44. House, Committee on Government Operations, Military Operations Subcommittee, *Report: Satellite Communications,* 89–96.
45. "Five Technological Satellites Will Be Developed by Hughes," *New York Times*, March 4, 1964, 7; "Hughes Gets ATS Pact," *Missiles and Rockets*, March 9, 1964, 8; D. D. Williams to Noel B. Hammond, "Information Related to PD-4286, Velocity Control, and Orientation of a Spin-Stabilized Body," July 27, 1964, HAC.
46. NASA, news release, "NASA Syncom Satellites Cross Paths 22,000 Miles above Pacific," #64-217, September 2, 1964, NHO; H. A. Rosen to L. A. Hyland, "Syncom Personnel," September 23, 1964, HAC.
47. "Protests Leveled on COMSAT Olympic Coverage," *Space Daily*, September 14, 1964, 237; A. S. Jerrems to H. A. Rosen, "Syncom Publicity," September 21, 1964, HAC; Anthony Michael Tedeschi, *Live via Satellite* (Washington, D.C.: Acropolis Books, 1989), 31.
48. NASA, news release, "NASA to Transfer Syncom Satellites to Defense Department," #65-5, December 31, 1964; DoD (PA), news release, "NASA to Transfer Syncom Satellites to Defense Department," January 4, 1965; DoD (PA), news release, "Syncom Communications Satellites Transfer Complete," #451-65, July 6, 1965.
49. Alexander R. Hammer, "Satellite Corp. Picks Six Candidates for New Board," *New York Times*, July 9, 1964, 43C, 45C; "COMSAT Achieves a New High at 48," *New York Times*, August 20, 1964; "COMSAT Stock Widely Distributed," *Space Daily*, August 14, 1964, 227; Goodman, "130,000 Owners Listed"; Finney, "COMSAT Takes On Private Role"; "Johnson Names Three to Fill COMSAT Board," *Aviation Week*, September 28, 1964, 26.
50. House, Committee on Government Operations, Military Operations Subcommittee, *Report: Satellite Communications*, 105–13.
51. "COMSAT Says Pentagon Ruined Plan," *Louisville (Ky.) Courier-Journal*, August 12, 1964; Comsat, press release, January 25, 1965, NHO; "COMSAT's Defense Bid Challenged by Philco, " *Washington Evening Star*, February 2, 1965, 3; "FCC Bars COMSAT Pact," *Missiles and Rockets*, February 8, 1965, 9; "COMSAT to Seek Bids on DoD COMSAT," *Space Daily*, February 17, 1965, 239–40; Larry Weekley, "COMSAT Bows to FCC, Invites General Bids," *Washington Post,* February 17, 1965.
52. Robert C. Cowen, "Commercial Relay Satellite Has Date for Spring," *Christian Science Monitor*, December 15, 1964, 3; Eric Wentworth, "COMSAT Gyrations Scrutinized by Federal Agencies in Case a Sudden Price Plunge Sparks Public Furor," *Wall Street Journal*, December 22, 1964, 3; "NASA Signs Agreement to Launch COMSAT's Early Bird," *Space Daily*, December 28, 1964, 269.
53. W. J. Weber, "Memorandum for the Record: Summary of HS-303, 'Early Bird' Communications Satellite Program," January 15, 1965, NHO.
54. Comsat, press release, "Early Bird Fact Sheet I: Key Early Bird Project Officials," March 30, 1965; NASA, "HS-303 Early Bird Communications Satellite," NASA Mission Operation Report S-631-65-01, March 31, 1965, NHO.
55. NASA, "HS-303 Early Bird Communications Satellite."
56. Ibid.; Comsat, press release, "Early Bird Fact Sheet III: Early Bird," March 30, 1965.

57. NASA, "Early Bird I Post-Launch Report No. 1," NASA MOR S-631-65-01, April 7, 1965, NHO; NASA, "Early Bird I Post-Launch Report No. 2," NASA MOR S-631-65-01, April 16, 1965, NHO.
58. Barry Miller, "Hughes Proposes TV Broadcast Satellite," *Aviation Week*, February 1, 1965, 75–77; "Early Bird Launched into Orbit," *Baltimore Sun*, April 7, 1965; Howard Simon, "Test of Early Bird's TV Heightens Optimism," *Washington Post*, April 8, 1965; "Early Bird Satellite Relays a TV Signal in a Surprise Test," *Wall Street Journal*, April 8, 1965; "Early Bird Orbit Is Nearly Perfect," *New York Times*, April 10, 1965, 11; "COMSAT Fixes Bird Orbit," *Baltimore Sun*, April 10, 1965; Suzanne Montgomery, "Early Bird Operation May be Speeded," *Missiles and Rockets*, April 12, 1965, 12; "As COMSAT Gets Down to Business," *U.S. News & World Report*, April 12, 1965, 7; Val Adams, "Two Continents See Global TV Today," *New York Times*, May 2, 1965; Jack Gould, "COMSAT to Assay ABC Satellite," *New York Times*, May 15, 1965, 63; COMSAT, press release, May 26, 1965.
59. "First Formal Bid to Be COMSAT Customer Filed by AT&T with FCC," *Wall Street Journal*, June 3, 1965, 4; "RCA Follows AT&T in Asking to Become Customer of COMSAT," *Wall Street Journal*, June 4, 1965, 6; Larry Weekley, "Early Bird Line Demand Grows," *Washington Post*, June 8, 1965, D8; "COMSAT Gets Go-Ahead to Put Early Bird into Commercial Use," *Wall Street Journal*, June 21, 1965, 3; "Early Bird Given Leasing Go-Ahead," *New York Times*, June 24, 1965; Jerry E. Bishop, "Lagging Early Bird," *Wall Street Journal*, August 2, 1965, 1, 14.
60. COMSAT, press release, "COMSAT Shareholders to Elect Twelve Directors at May 11 Meeting," May 7, 1965; COMSAT, press release, "Welch Announces Wish to Retire as COMSAT Chief Executive Officer and Board Chairman," July 7, 1965; COMSAT, press release, "COMSAT Board Elects James McCormack Chief Executive Officer and Board Chairman," October 15, 1965; Gene Smith, "COMSAT Names a New Chairman," *New York Times*, October 16, 1965, 30.
61. Deputy Associate Administrator (Hilburn) to Associate Administrator (Seamans), "Possible Use of Early Bird in Support of the Apollo Network," July 19, 1965, NHO; William R. Corliss, "History of the Goddard Networks," November 1, 1969, NHO.
62. COMSAT, press release, "COMSAT Files Application for FCC Authorization for Proposed Satellites for Apollo Service," September 30, 1965; "COMSAT Seeks to Buy Four Super Early Birds," *New York Times*, October 20, 1965, 11; "COMSAT Files Apollo Satellite Contract," *Space Daily*, October 21, 1965, 262; "First Commercial Satellite to Be Placed over Pacific," *New York Times*, November 4, 1965, 28C; Robert C. Seamans Jr. to Dr. Joseph V. Charyk, November 23, 1965, NHO; "Four Satellite System Ordered by COMSAT," *New York Times*, November 25, 1965, 3; Robert C. Seamans Jr. to Dr. Joseph V. Charyk, December 30, 1965, NHO.
63. "COMSAT Studies," *Aviation Week & Space Technology*, April 6, 1964, 19; "The Big Four in COMSAT Competition," *Space Daily*, April 9, 1965, 223; "Early Bird Follow-On," *Aviation Week & Space Technology*, August 23, 1965, 27; Lyle Denniston, "COMSAT Satellite Decision Hints of Earlier Profits," *Washington Evening Star*, May 11, 1965, 72.

64. COMSAT, press release, "COMSAT Asks Manufacturers to Submit Proposals on Advanced Satellite for Global Communications," August 17, 1965; "Early Bird Follow-On," *Aviation Week & Space Technology*, August 23, 1965, 27; "Telephone Satellite Predicted for 1968," *Miami Herald*, December 3, 1965; Larry Weekley, "COMSAT Negotiates with TRW on Large Satellite Contract," *Washington Post*, December 17, 1965, C8; "COMSAT Is Negotiating for New Satellites with TRW; Users' Charges May Be Cut," *Wall Street Journal*, December 17, 1965, 24.
65. John Noble Wilford, "COMSAT Seeking Bigger Satellite," *New York Times*, December 30, 1965, 21; "COMSAT Asks Designs for Big, New Satellite," *Washington Evening Star*, December 30, 1965, A9.
66. COMSAT, press release, "COMSAT Gives Highlights of Corporation's Progress in an Interim Report to Shareholders," December 31, 1965.

Chapter 7. Outcomes

1. J. R. Pierce to Eugene M. Emme, January 7, 1966, NHO.
2. Correspondence between John R. Pierce and the author, 1996.
3. The author's notes of the two conferences are the source for the following material.
4. Discussions and correspondence with the author, 1995–96.
5. *NASA Historical Data Book*, 6 vols. (Washington, D.C.: GPO, 1988–2000), 2:222, 3:255.
6. More than half of this was AT&T's original Telstar investment. The second-largest R&D investor was Comsat. Hughes was probably third, but all of the manufacturers spent substantial amounts of money on R&D. This became a very competitive business where low price *and* high performance were required.
7. The largest portion of this was the cost of the fourteen Atlas-Centaur-launched satellites of the fourth generation. Intelsat spent almost one-half billion dollars on the *Intelsat 4* and *Intelsat 4A* satellites alone. See Emeric Podraczky and Joseph N. Pelton, "Intelsat Satellites," in *The INTELSAT Global Satellite System*, ed. Joel Alper and Joseph N. Pelton, AIAA Progress in Astronautics and Aeronautics, vol. 93 (New York: American Institute of Aeronautics and Astronautics, 1984), 112.
8. Hughes sold nine Delta-launched HS-333 satellites to Western Union, Telesat Canada, and Indonesia in the mid-1970s.
9. Strangely enough, Intelsat later chose to emphasize 72-MHz transponders for all but global beams.
10. This statement has been made to the author and to other Lockheed Martin Global Telecommunications employees.
11. The export-control regime implemented in 1999 as a result of the Cox report—which alleged that U.S. satellite manufacturers transferred ICBM technology (perhaps inadvertently) to China in the course of launching U.S.-made satellites in Chinese rockets—may give European companies the chance to dominate satellite manufacturing just as they already dominate launch services.

Glossary and Acronyms

Agena. An upper-stage rocket designed to provide in-orbit propulsion for low-altitude spy satellites.

AKM (Apogee Kick Motor). A rocket stage, typically weighing as much as the rest of the satellite, which circularizes the highly eccentric geosynchronous transfer orbit (GTO).

AM (Amplitude Modulation). A method of transmitting information by varying the amplitude of the carrier signal in proportion to variations in amplitude of the information signal. This usually results in two sidebands, one above and one below the carrier frequency. A simple 4-kHz voice signal modulated onto a 20-kHz carrier would have sidebands at 16 kHz and 24 kHz.

Apogee Kick Motor (AKM). The rocket engine, usually attached to the satellite rather than part of the launch vehicle, that is fired at the apogee of the geosynchronous transfer orbit (GTO) to inject the satellite into geostationary orbit (GSO).

Atlas. The first U.S. ICBM. An Air Force project, it grew from the MX-774 project of the 1940s but was not given priority until the mid-1950s. It was operational only until the solid-propulsion Minuteman was deployed. It required considerable preparation before launch.

Atlas-Agena. The Atlas first stage with an Agena mounted on top.

Atlas-Centaur. The Atlas first stage with a Centaur mounted on top.

Bandwidth. A range of frequencies. A modulated carrier will have a bandwidth related to the frequency of the message signal. A simple 4-kHz voice signal amplitude modulated onto a 20-kHz carrier would have a 3-dB (half-power) bandwidth of 8 kHz ranging from 16 kHz to 24 kHz.

Carrier-to-Noise Ratio (C/N). The ration of carrier energy to noise energy in a signal.

C-band. The frequencies from 3.9 to 6.2 GHz. In satellite communications (FSS), C-band is used to refer to downlink frequencies between 3.4 GHz and 4.2 GHz and uplink frequencies between 5.85 GHz and 7.075 GHz. It is often referred to as 4/6 GHz.

CCIR (International Consultative Committee on Radio). Now ITU-R.

Centaur. An upper-stage rocket using cryogenic fuels: liquid hydrogen and liquid oxygen.

CEPT (Conference of European Postal and Telecommunications Administrations). European bloc that challenged the U.S. plans to dominate satellite communications.

C/N. *See* Carrier-to-Noise Ratio.

Common Carrier. A company providing carriage service to the common market, rather than to a restricted market.

Comsat (Communications Satellite Corporation). Formed by act of Congress to operate communications satellites.

dB (decibel). One tenth of a Bell. 10*log(ratio).

Delta. One of the most reliable satellite launch vehicles. It was derived from the Thor IRBM first stage and refinements of the Vanguard second and third stages. From its introduction in 1960 to the present, it has been approximately 95 percent successful.

DoD (Department of Defense).

Downlink. The communications signal transmitted from the satellite down to the ground.

Earth Station. In satellite communications, the ground station with antenna and related electronics.

EIRP (Effective Isotropic Radiated Power; Equivalent Isotropic Radiated Power). The power required to produce the same energy density if the transmission pattern were isotropic (omni-directional). Think of the ability of a 2-watt flashlight bulb two inches away to adequately illuminate a small portion of a newspaper, as compared with a 100-watt lightbulb fifteen feet away.

FCC (Federal Communications Commission). The U.S. regulatory agency responsible for both wire and nonwire (radio) communications.

FM (Frequency Modulation). A method of transmitting information by varying the frequency of the carrier signal in proportion to amplitude variations of the information signal.

FSS (Fixed Satellite Service).

Gain. In an amplifier, the ratio of the output signal to the input signal. The gain of a transmitting antenna is a measure of the antenna's ability to focus the available energy in one direction. Transmitting gain is proportional to the square of the antenna diameter divided by the carrier wavelength = $(D/l)^2$. Receiving gain of an antenna is a function of area—and is therefore proportional to the square of the diameter of the antenna. It is convenient to divide this number by the square of the carrier wavelength such that transmit and receive gain are equal.

GEO (Geosynchronous Earth Orbit). A twenty-four-hour orbit (actually twenty-three hours and fifty-six minutes). A satellite in 42,165-kilometer orbit (35,487 kilometers above the surface of Earth) around Earth will have this period.

GHz (GigaHertz). One billion cycles per second.

GSO (Geostationary Orbit). GEO with zero inclination and zero eccentricity.

G/T (Gain/Noise Temperature). The measure of a communications receive system to receive information with minimum errors. It is a function of antenna diameter and the thermal noise level of the first stage of amplification (low-noise amplifier).

GTO (Geosynchronous Transfer Orbit). An orbit with a low perigee and a GEO apogee.

Hertz. A measure of frequency—after Heinrich Hertz.

Hypergolic. Fuels that spontaneously ignite when brought together.

ICBM (Intercontinental Ballistic Missile). A war rocket with a range on the order of 10,000 kilometers.

Intelsat (International Telecommunications Satellite Organization). Formed in 1964 to own and operate the global communications satellite system.

IRBM (Intermediate Range Ballistic Missile). A war rocket with a range on the order of 2,000 kilometers.

JPL (Jet Propulsion Laboratory). A government-funded rocket lab at CalTech. The JPL switched its emphasis from rockets to planetary exploration after the formation of NASA.

Ka-band. The frequencies from 33 to 36 GHz. In satellite communications (FSS), Ka-band is used to refer to downlink frequencies between 18.6 GHz and 21.2 GHz and uplink frequencies between 27 GHz and 31 GHz. It is often referred to as 20/30 GHz.

LEO (Low Earth Orbit). Typically less than 1,500 kilometers above Earth's surface; also below the lower van Allen radiation belt.

MASER (Microwave Amplification by Stimulated Emission of Radiation). A very low noise amplifier.

MEO (Medium Earth Orbit). An orbit around 10,000 kilometers above Earth's surface, between the two van Allen belts.

MHz (MegaHertz). A measure of frequency: one million cycles per second.

Microwave. Radio frequency (RF) carrier waves with wavelengths of less than one meter—frequencies above 300 MHz. "Microwave" is typically used to refer to frequencies above 1 GHz, but it nominally includes all of UHF.

Modulation. The process by which the characteristics of a carrier wave are varied in accordance with a message signal (voice, data, video).

Multiple Access. The ability to transmit more than one signal through a transponder. Multiple access requires a highly linear amplifier.

NASA (National Aeronautics and Space Administration). An independent agency of the U.S. government, dedicated to air and space research and development.

PCM (Pulse Code Modulation). A type of digital modulation in which the information is carried by "pulses."

RFP (Request For Proposals). The specifications and other documents by which a buying organization invites proposals and price quotations.

Sideband. The signals, to either side of a main carrier, that carry the information in AM transmission.

SSB (Single Sideband). A form of AM transmission in which only one sideband is transmitted—probably the most frequency-efficient form of modulation.

Telephony. The transmission of voice signals.

Television. The transmission of video signals.

Thruster. A small rocket used to station-keep a GEO satellite. Large thrusters generate 20 Newtons (five pounds) of thrust. Compare this to the thousands and millions of Newtons generated by launch vehicle main engines.

Transponder. A microwave repeater; the "channels" of a communications satellite. Each transponder can retransmit a signal or set of signals at a given power (EIRP).

Traveling-Wave Tube (TWT). An electronic signal amplifier characterized by wide bandwidth, high gain, and linearity.

Traveling-Wave Tube Amplifier (TWTA). The main transmitters (microwave repeaters) on a satellite. One of these is associated with each transponder and determines the available RF communications power.

TT&C (Telemetry, Tracking, and Command). The equipment and systems used to control a satellite.

TWT. See Traveling-Wave Tube.

Uplink. In satellite communications, the signal from the Earth station to the space station (satellite).

UTC (Universal Time Coordinated). The very precise time used at astronomical observatories and global enterprises. Roughly equivalent to Greenwich (England) Mean Time (GMT).

WARC (World Administrative Radio Conference). The ITU meeting at which decisions regarding the global use of communications, especially radio communications, are made.

Wavelength. The distance between successive peaks or troughs. The wavelength is related to the frequency by the formula $\mu = \lambda f$, where μ is wave velocity, λ is wavelength, and f is frequency.

X-band. The frequencies from 5.2 to 10.9 GHz. In satellite communications (FSS), X-band is used to refer to downlink frequencies between 7.25 GHz and 7.75 GHz and uplink frequencies between 7.9 GHz and 8.4 GHz. It is used extensively by the military.

Bibliography

Note on Sources

The sources used in this work include books, published articles, government documents, and unpublished materials. In some cases, the source of an unpublished document (e.g., press releases) is obvious; in other cases, the following are used as identifiers:

AT&T: AT&T Archives, Warren, N.J.

HAC: Hughes Aircraft Company Archives, El Segundo, Calif.

JHR: Personal files of John H. Rubel, Tesuque, N.M.

NAII: National Archives II, College Park, Md.

NHO: NASA History Office, Washington, D.C.

Many of the documents from the AT&T and Hughes Archives, and the documents from John H. Rubel, were obtained through the kindness of Helen M. Gavaghan, who shared her research with me. No formal oral histories or interviews were performed. However, over the years I have spoken informally to many of the participants, and we have exchanged letters. Some of this personal correspondence is now in my files.

Books

Aitken, Hugh G. J. *The Continuous Wave: Technology and American Radio, 1900–1932.* Princeton: Princeton University Press, 1985.

Alic, John A., Lewis M. Branscomb, Harvey Brooks, Ashton B. Carter, and Gerald L. Epstein. *Beyond Spinoff: Military and Commercial Technologies in a Changing World.* Cambridge: Harvard University Business School Press, 1992.

Allen, Frederick Lewis. *The Big Change: America Transforms Itself, 1900–1950.* 1952. First Perennial Library ed., New York: Harper and Row, 1986.

Alper, Joel, and Joseph N. Pelton, eds. *The INTELSAT Global Satellite System*. Progress in Astronautics and Aeronautics, vol. 93. New York: American Institute of Aeronautics and Astronautics, 1984.

Baldwin, Thomas F., and D. Stevens McVoy. *Cable Communications*. Englewood Cliffs, N.J.: Prentice-Hall, 1983.

Barnouw, Erik. *Tube of Plenty: The Evolution of American Television*. Rev. ed. Oxford: Oxford University Press, 1982.

Bauer, Eugene E. *Boeing in Peace and War*. Enumclaw, Wash.: TABA, 1990.

Biddle, Wayne. *Barons of the Sky*. New York: Simon and Schuster, 1991.

Bulkeley, Rip. *The Sputniks Crisis and the Early United States Space Policy*. Bloomington: Indiana University Press, 1991.

Byerly, Radford, Jr., ed. *Space Policy Reconsidered*. Boulder, Colo.: Westview Press, 1989.

Clarke, Arthur C. *The Exploration of Space*. New York: Harper and Row, 1952.

———. *How the World Was One*. New York: Bantam, 1992.

Cohen, Linda R., and Roger G. Noll, eds. *The Technology Pork Barrel*. Washington, D.C.: Brookings Institution, 1991.

Davies, Merton E., and William R. Harris. *RAND's Role in the Evolution of Balloon and Satellite Observation Systems and Related U.S. Space Technology*. Santa Monica, Calif.: RAND, 1988.

DeVorkin, David H. *Science with a Vengeance*. New York: Springer-Verlag, 1992.

Elder, Donald C. *Out from behind the Eight-Ball: A History of Project Echo*. San Diego: American Astronautical Society, 1995.

Emme, Eugene, ed. *The History of Rocket Technology*. Detroit: Wayne State University Press, 1964.

Galloway, Jonathan F. *The Politics and Technology of Satellite Communications*. Lexington, Mass.: Lexington Books, 1972.

Glennan, T. Keith. *The Birth of NASA: The Diary of T. Keith Glennan*. Edited by J. D. Hunley. Washington, D.C.: NASA History Office, 1993.

Gorn, Michael H. *Harnessing the Genie: Science and Technology Forecasting for the Air Force, 1944–1986*. Washington, D.C.: Office of Air Force History, U.S. Air Force, 1988.

Gromyko, Andrei, to Central Committee CPSU. May 20, 1961. Quoted in Aleksandr Fursenko and Timothy Naftali, *One Hell of a Gamble: Khrushchev, Castro, and Kennedy, 1958–1964*, 120–21. New York: Norton, 1997.

Guile, Bruce R., and Harvey Brooks, eds. *Technology and Global Industry*. Washington, D.C.: National Academy Press, 1987.

Hyland, L. A. *Call Me Pat*. Virginia Beach, Va.: Donnington, 1993.

Kinsley, Michael E. *Outer Space and Inner Sanctums*. New York: John Wiley and Sons, 1976.

Kistiakowsky, George B. *A Scientist at the White House*. Cambridge: Harvard University Press, 1976.

Kofsky, Frank. *Harry S. Truman and the War Scare of 1948*. New York: St. Martin's Press, 1995.

Koppes, Clayton R. *JPL and the American Space Program: A History of the Jet Propulsion Laboratory.* New Haven: Yale University Press, 1982.
Logsdon, John M. *The Decision to Go to the Moon: Project Apollo and the National Interest.* Cambridge: MIT Press, 1970.
Mack, Pamela E. *Viewing the Earth: The Social Construction of the Landsat Satellite System.* Cambridge: MIT Press, 1990.
Magnant, Robert S. *Domestic Satellite: An FCC Giant Step.* Boulder, Colo.: Westview Press, 1977.
McCullough, David. *Truman.* New York: Simon and Schuster, 1992.
McDougall, Walter A. *The Heavens and the Earth: A Political History of the Space Age.* New York: Basic Books, 1985.
Medaris, John B. *Countdown for Decision.* New York: Putnam, 1960.
Milward, Alan S. *War, Economy, and Society, 1939–1945.* Berkeley: University of California Press, 1977.
Mowery, David C., and Nathan Rosenberg. *Technology and the Pursuit of Economic Growth.* Cambridge: Cambridge University Press, 1989.
NASA Historical Data Book. 6 vols. Washington, D.C.: GPO, 1988–2000.
NASA. *Aeronautical and Astronautical Events of 1961.* Washington, D.C.: GPO, 1962.
———. *Final Report on the Relay 1 Program.* Washington, D.C.: GPO, 1965.
———. *Fourth Semi-Annual Report to Congress,* 10–17. Washington, D.C.: GPO, 1961.
———. *Proceedings of the First National Conference on the Peaceful Uses of Outer Space.* Washington, D.C.: GPO, 1961.
Naugle, John E. *First among Equals.* Washington, D.C.: GPO, 1991.
Nelson, Richard R., ed. *Government and Technical Progress.* New York: Pergamon Press, 1982.
———. *National Innovation Systems.* Oxford: Oxford University Press, 1993.
Neufeld, Jacob. *The Development of Ballistic Missiles in the United States Air Force, 1946–1960.* Washington, D.C.: Office of Air Force History, United States Air Force, 1990.
Neustadt, Richard E., and Ernest R. May. *Thinking in Time.* New York: Free Press, 1986.
Newell, Homer E. *Beyond the Atmosphere: Early Years of Space Science,* NASA SP-4211. Washington, D.C.: GPO, 1980.
Ordway, Frederick I., III, and Randy Liebermann, eds. *Blueprint for Space: Science Fiction to Science Fact.* Washington, D.C.: Smithsonian Institution Press, 1992.
Ordway, Frederick I., III, and Mitchell R. Sharpe. *The Rocket Team.* New York: Thomas Y. Crowell, 1979.
Otega y Gasset, José. *La Rebelión de la masas.* 1944. Reprint, Madrid: Espasa-Calpe, 1976.
Pelton, Joseph N., and Marcellus S. Snow, eds. *Economic and Policy Problems in Satellite Communications.* New York: Praeger, 1977.
Pierce, John R. *The Beginnings of Satellite Communications.* San Francisco: San Francisco Press, 1968.
Rhodes, Richard. *The Making of the Atomic Bomb.* New York: Simon and Schuster, 1986.

Roland, Alex, ed. *A Spacefaring People: Perspectives on Early Space Flight.* Washington, D.C.: Scientific and Technical Information Branch, NASA, 1985.

Rosen, Milton W. *The Viking Rocket Story.* New York: Harper, 1955.

Rosenberg, Nathan. *Inside the Black Box: Technology and Economics.* Cambridge: Cambridge University Press, 1982.

Rosholt, Robert L. *An Administrative History of NASA, 1958–1963.* Washington, D.C.: GPO, 1966.

Russo, Arturo. *The Early Development of the Telecommunications Satellite Programme in ESRO, 1965–1971.* Noordwijk, The Netherlands: ESA, 1993.

Ryan, Cornelius. *Across the Space Frontier.* New York: Viking, 1952.

———. *Conquest of the Moon.* New York: Viking, 1953.

Scherer, F. M. *Innovation and Growth.* Cambridge: MIT Press, 1989.

Seamans, Robert C., Jr. *Aiming at Targets.* Washington, D.C.: NASA, 1996.

Smith, Delbert D. *Communication via Satellite: A Vision in Retrospect.* Boston: A. W. Sijthoff, 1976.

Staudenmaier, John M. *Technology's Storytellers: Reweaving the Human Fabric.* Cambridge: MIT Press, 1985.

Tedeschi, Anthony Michael. *Live via Satellite.* Washington, D.C.: Acropolis Books, 1989.

Van Dyke, Vernon. *Pride and Power: The Rationale of the Space Program.* Urbana: University of Illinois Press, 1964.

Von Braun, Wernher. *The Mars Project.* Translated by Henry J. White. Urbana: University of Illinois Press, 1953.

Wilhelm, Don. *Toward a Rocky Moon.* Tucson: University of Arizona Press, 1993.

Winter, Frank H. *Prelude to the Space Age: The Rocket Societies, 1924–1940.* Washington, D.C.: Smithsonian Institution Press, 1983.

York, Herbert. *The Advisors: Oppenheimer, Teller, and the Superbomb.* San Francisco: Freeman, 1976.

———. *Making Weapons, Talking Peace.* New York: Basic Books, 1987.

Published Articles

Adams, Val. "Two Continents See Global TV Today." *New York Times,* May 2, 1965, 35.

Allen, Robert S., and Paul Scott. "Satellite Corp. Growth Pains." *Newport News (Va.) Times Herald,* January 4, 1963, 10.

Andrews, Edmund L. "Big Award Is Ordered for Hughes." *New York Times,* August 18, 1995, D5.

"As COMSAT Gets Down to Business." *U.S. News & World Report,* April 12, 1965, 7.

"The Big Four in COMSAT Competition." *Space Daily,* April 9, 1965, 223.

Bishop, Jerry E. "Lagging Early Bird." *Wall Street Journal,* August 2, 1965, 1, 14.

Brownlow, Cecil. "International Comsat Agency Considered." *Aviation Week & Space Technology,* February 17, 1964, 34.

Brownlow, Cecil. "U.S.-Europe Comsat Agreement Predicted." *Aviation Week & Space Technology,* April 22, 1963, 74–75.

Cantelon, Philip L. "The Origins of Microwave Telephony." *Technology and Culture,* July 1995, 560–82.

Charyk, Joseph V. "Communications Satellite Corporation: Objectives and Problems." *Astronautics and Aerospace Engineering,* September 1963, 45–47.

Clarke, Arthur C. "Extra-Terrestrial Relays." *Wireless World* 51, no. 10 (October 1945): 303–8.

Cohen, Linda R., and Roger G. Noll. "The Applications Technology Satellite Program." In *The Technology Pork Barrel,* edited by Linda R. Cohen and Roger G. Noll, 149–77. Washington, D.C.: Brookings Institution, 1991.

Colino, Richard R. "The INTELSAT System: An Overview." In *The INTELSAT Global Satellite System,* edited by Joel Alper and Joseph N. Pelton, 55–94. Progress in Astronautics and Aeronautics, vol. 93. New York: American Institute of Aeronautics and Astronautics, 1984.

"Communications Satellite Corp. Gets Bids for Designing System from Six Companies." *Wall Street Journal,* February 12, 1964, 1.

"Communications Satellite Firm Negotiates Spacecraft Contract with Hughes Aircraft." *Wall Street Journal,* March 17, 1964, 6.

"COMSAT Achieves a New High at 48." *New York Times,* August 20, 1964, 37.

"COMSAT Asks Designs for Big, New Satellite." *Washington Evening Star,* December 30, 1965, A9.

"COMSAT Asks to Build Puerto Rico Station." *Washington, D.C., Sunday Star,* March 13, 1966, E11.

"COMSAT Files Apollo Satellite Contract." *Space Daily,* October 21, 1965, 262.

"COMSAT Files with FCC." *Missiles and Rockets,* March 9, 1964, 8.

"COMSAT Fixes Bird Orbit." *Baltimore Sun,* April 10, 1965.

"COMSAT Gets Go-Ahead to Put Early Bird into Commercial Use." *Wall Street Journal,* June 21, 1965, 3.

"COMSAT Is Negotiating for New Satellites with TRW; Users' Charges May Be Cut." *Wall Street Journal,* December 17, 1965, 24.

"COMSAT Launches Pacific Satellite." *New York Times,* October 27, 1966, 57.

"COMSAT Making Bid for New TV Satellite." *Washington Evening Star,* February 5, 1966, A5.

"COMSAT Poised for Act II in Space Satellites." *Washington, D.C., Sunday Star,* January 9, 1966, R22.

"The COMSAT Problem." *Aviation Week & Space Technology,* April 6, 1964, 11.

"COMSAT Reports Initial Commercial Revenue of $2,139,000 in 1965." *Wall Street Journal,* April 1, 1966, 11.

"COMSAT Says Pentagon Ruined Plan." *Louisville (Ky.) Courier-Journal,* August 12, 1964.

"COMSAT Seeks Studies on Multi-Role Vehicle." *Aviation Week & Space Technology,* January 3, 1966, 21.

"COMSAT Seeks to Buy Four Super Early Birds." *New York Times,* October 20, 1965, 11.
"COMSAT Static." *Aviation Week & Space Technology,* March 30, 1964, 15.
"COMSAT Stock Widely Distributed." *Space Daily,* August 14, 1964, 277.
"COMSAT Studies." *Aviation Week & Space Technology,* April 6, 1964, 19.
"COMSAT System Selection Pending." *Space Daily,* March 25, 1964, 463.
"COMSAT to Seek Bids on DoD COMSAT." *Space Daily,* February 17, 1965, 239–40.
"COMSAT Will Seek Puerto Rico Station." *New York Times,* March 12, 1966, 22.
"COMSAT's Defense Bid Challenged by Philco." *Washington Evening Star,* February 2, 1965, 3.
Cowen, Robert C. "Commercial Relay Satellite Has Date for Spring." *Christian Science Monitor,* December 15, 1964, 3.
Denniston, Lyle. "COMSAT Satellite Decision Hints of Earlier Profits." *Washington Evening Star,* May 11, 1965, 72.
"Development of the Relay Communications Satellite." *Interavia,* June 1962, 758–59.
"Early Bird Earns Back 20 Cents on Every $1." *Washington Evening Star,* April 1, 1966, B8.
"Early Bird Follow-On." *Aviation Week & Space Technology,* August 23, 1965, 27.
"Early Bird Given Leasing Go-Ahead." *New York Times,* June 24, 1965, 57.
"Early Bird Income Put at $2.1 Million." *Washington Post,* April 1, 1966, 10.
"Early Bird Launched into Orbit." *Baltimore Sun,* April 7, 1965.
"Early Bird Orbit Is Nearly Perfect." *New York Times,* April 10, 1965, 11.
"Early Bird Put in Reserve; Better Satellite Takes Over." *Philadelphia Sunday Bulletin,* February 2, 1969, 7.
"Early Bird Satellite Relays a TV Signal in a Surprise Test." *Wall Street Journal,* April 8, 1965, 2.
Engstrom, Elmer W. "24-Hour Communications Satellite Systems." In NASA, *Proceedings of the First National Conference on the Peaceful Uses of Outer Space,* 139–42. Washington, D.C.: GPO, 1961.
Ergas, Henry. "Does Technology Policy Matter?" In *Technology and Global Industry,* edited by Bruce R. Guile and Harvey Brooks, 191–245. Washington, D.C.: National Academy Press, 1987.
"FCC Bars COMSAT Pact." *Missiles and Rockets,* February 8, 1965, 9.
"FCC Gives ComSat Go-Ahead." *Missiles and Rockets,* April 20, 1964, 10.
"FCC Would Allow All Phone Firms to Invest in Space Message Firm." *Wall Street Journal,* November 29, 1962.
Finney, John W. "COMSAT Takes On Private Role as Holders Stage First Meeting." *New York Times,* September 18, 1964, 45.
———. "Space Radio Plan Creating Doubts." *New York Times,* April 24, 1963, 2.
———. "Speed-Up Urged on Three Satellites." *New York Times,* March 31, 1960, 9.
"First Commercial Satellite to Be Placed over Pacific." *New York Times,* November 4, 1965, 28C.
"First Formal Bid to Be COMSAT Customer Filed by AT&T with FCC." *Wall Street Journal,* June 3, 1965, 4.

"Five Technological Satellites Will Be Developed by Hughes." *New York Times,* March 4, 1964, 7.

"Four Satellite System Ordered by COMSAT." *New York Times,* November 25, 1965, 3.

Goodman, S. Oliver. "130,000 Owners Listed in Initial Report of COMSAT." *Washington Post,* August 14, 1964, D6.

Gould, Jack. "COMSAT to Assay ABC Satellite." *New York Times,* May 15, 1965, 63.

Hagen, John P. "Viking and Vanguard." In *The History of Rocket Technology,* edited by Eugene Emme, 122–27. Detroit: Wayne State University Press, 1964.

Hall, R. Cargill. "Early U.S. Satellite Proposals." In *The History of Rocket Technology,* edited by Eugene Emme, 68. Detroit: Wayne State University Press, 1964.

Hammer, Alexander R. "Satellite Corp. Picks Six Candidates for New Board." *New York Times,* July 9, 1964, 43C, 45C.

Haviland, R. P. "The Communication Satellite." In *Eighth International Astronautical Congress Proceedings*, 543–62. Vienna: Springer-Verlag, 1958.

Hotz, Robert. "Selling Shares in Space." *Aviation Week & Space Technology*, May 11, 1964, 17.

"House Group Studies NASA Role as COMSAT Corp. Technical Adviser." *Aviation Week & Space Technology,* April 13, 1964, 32.

"House Votes 354-to-9 Approval of Space Communications Firm." *Washington Post,* May 4, 1962, A1.

"Hughes Gets ATS Pact." *Missiles and Rockets,* March 9, 1964, 8.

"Incentive Potential of $4.8 Million in Hughes-COMSAT Corp. Contract." *Aviation Week & Space Technology,* March 30, 1964.

"Johnson Names Three to Fill COMSAT Board." *Aviation Week & Space Technology,* September 28, 1964, 26.

Klass, Philip J. "Civil Communication Satellites Studied." *Aviation Week,* June 22, 1959, 189–97.

———. "COMSAT Firm Files $200-Million Fund Plan." *Aviation Week & Space Technology,* May 11, 1964, 25–26.

Launius, Roger A. Introduction to *The Birth of NASA: The Diary of T. Keith Glennan,* edited by J. D. Hunley, x–xi. Washington, D.C.: NASA History Office, 1993.

Liebermann, Randy. "The Collier's and Disney Series." In *Blueprint for Space: Science Fiction to Science Fact,* edited by Frederick I. Ordway III and Randy Liebermann, 135–46. Washington, D.C.: Smithsonian Institution Press, 1992.

Lutz, Samuel G. "Satellite Communications and Frequency Sharing." *Interavia*, June 1962, 753–57.

Mack, Pamela. "Satellites and Politics: Weather, Communications, and Earth Resources." In *A Spacefaring People: Perspectives on Early Space Flight,* edited by Alex Roland, 32–38. Washington, D.C.: Scientific and Technical Information Branch, NASA, 1985.

MacKenzie, John P. "Industry Snaps Up Its Half of COMSAT Stock." *Washington Post,* March 28, 1964, 2.

———. "Two Hundred Carriers File for COMSAT Shares." *Washington Post,* March 24, 1964, 4.

MacKenzie, John P., Sidney Metzger, and Robert H. Pickard. "Relay." *Aeronautics and Aerospace Engineering,* September 1963, 64–67.

Miller, Barry. "Hughes Proposes TV Broadcast Satellite." *Aviation Week,* February 1, 1965, 75–77.

Montgomery, Suzanne. "COMSAT Board Awaits Senate Nod." *Missiles and Rockets,* September 28, 1964, 34.

———. "Early Bird Operation May Be Speeded." *Missiles and Rockets,* April 12, 1965, 12.

"NASA's ComSat Funding to Climb." *Missiles and Rockets,* April 2, 1962, 17.

"NASA Sees Telstar-Type Satellite as Best for World-Wide System." *Aviation Week & Space Technology,* September 24, 1962, 40.

"NASA Signs Agreement to Launch COMSAT's Early Bird." *Space Daily,* December 28, 1964, 269.

"Navy Prefers MA/GG COMSAT; COMSAT Act Repeal Urged by Senators." *Space Daily,* March 26, 1964, 471–72.

Norsell, Paul E. "SYNCOM." *Astronautics and Aerospace Engineering,* September 1963, 76–78.

"Observatory Pinpoints Lost Satellite Definitely." *Washington, D.C., Sunday Star,* March 3, 1963.

"Ocean Satellite by 1965 Is Sought." *New York Times,* March 5, 1964, 4.

Oslund, Jack. "Open Shores to Open Skies: Sources and Directions of U.S. Satellite Policy." In *Economic and Policy Problems in Satellite Communications,* edited by Joseph N. Pelton and Marcellus S. Snow, 181. New York: Praeger, 1977.

Pearson, Drew. "The Revolution in Communications." *Washington Post,* November 26, 1962, B23.

———. " Satellite Corp. Meets in Secrecy." *Washington Post,* November 27, 1962.

"Permission Sought to Orbit Communications Device in '65." *Baltimore Sun,* March 5, 1964.

Pierce, John R. "Orbital Radio Relays." *Jet Propulsion,* April 1955, 44.

Pierce, John R., and Rudolf Kompfner. "Transoceanic Communications by Means of Satellites." *Proceedings of the IRE,* March 1959, 372–80.

Podraczky, Emeric, and Joseph N. Pelton. "Intelsat Satellites." In *The INTELSAT Global Satellite System,* edited by Joel Alper and Joseph N. Pelton, 95, 133. Progress in Astronautics and Aeronautics, vol. 93. New York: American Institute of Aeronautics and Astronautics, 1984.

"The Poor and the Rich." *The Economist,* May 25, 1996, 24.

Porter, Frank C. "Satellite Company Incorporators Get Briefing from U.S." *Washington Post,* October 23, 1962.

"Protests Leveled on COMSAT Olympic Coverage." *Space Daily,* September 14, 1964, 237.

Raymond, Jack. "President Names Satellite Board." *New York Times,* October 5, 1962, 1.

"RCA Follows AT&T in Asking to Become Customer of COMSAT." *Wall Street Journal,* June 4, 1965, 6.

Reiger, Siegfried H. "Commercial Satellite Systems." *Astronautics and Aerospace Engineering,* September 1963, 26–30.

Rosen, Harold A. "Harold Rosen on Satellite Technology Then and Now." *Via Satellite,* July 1993, 40–43.

Rubinstein, Ellis. "Dollars vs. Satellites." *IEEE Spectrum,* October 1976, 75–80.

"Satellite Communication." *International Science and Technology,* January 1963, 69–74.

"Satellite Corp. Picks Six Candidates for New Board." *New York Times,* July 9, 1964, 43C.

"Satellite Corp., Prodded by FCC, May Tell More This Week about Plans for Stock Sale." *Wall Street Journal,* July 29, 1963, 7.

"Satellite Corp. to Market Stock at $20 a Share." *Wall Street Journal,* May 7, 1964.

"Satellite Firm Selects Top Officers, Hints at Stock Sale in a Year." *Wall Street Journal,* February 28, 1963, 32.

"Satellite Job Is Resigned by Graham." *Washington Evening Star,* January 26, 1963, A9.

"Satellite Urged for Domestic TV." *New York Times,* March 25, 1966, 48.

Scherer, F. M. "Invention and Innovation in the Watt-Boulton Steam Engine Venture." *Technology and Culture,* spring 1965, 165–87.

Sehlstedt, Albert, Jr. "Lani Bird Misses Proposed Orbit." *Baltimore Sun,* November 1, 1966.

Smith, Gene. "COMSAT Names a New Chairman." *New York Times,* October 16, 1965, 30.

Smith, Marcia S. "Civilian Space Applications: The Privatization Battleground." In *Space Policy Reconsidered,* edited by Radford Byerly Jr., 105–16. Boulder, Colo.: Westview Press, 1989.

"South African Observatory Spots Missing Syncom." *Washington Post,* March 1, 1963, A8.

"Space Projects Shifted." *New York Times,* September 23, 1960, 2.

Stuhlinger, Ernst. "Gathering Momentum." In *Blueprint for Space: Science Fiction to Science Fact,* edited by Frederick I. Ordway III and Randy Liebermann, 119. Washington, D.C.: Smithsonian Institution Press, 1992.

Taylor, Hal. "Council Favors Private Ownership." *Missiles and Rockets,* July 3, 1961, 11.

"Telephone Satellite Predicted for 1968." *Miami Herald,* December 3, 1965.

"$32 Million Order Filed by COMSAT." *Washington Evening Star,* April 29, 1966, F2.

Toth, Robert C. "F.C.C. Sees Delay in Space Stock." *New York Times,* July 27, 1963, 3.

"Two Senators Hit COMSAT, U.S. Deal." *Washington Post,* March 26, 1964, 7.

"U.S. Will Develop Radio Satellites." *New York Times,* August 29, 1959, 14.

Weekley, Larry. "COMSAT Bows to FCC, Invites General Bids." *Washington Post,* February 17, 1965.

———. "COMSAT Negotiates with TRW on Large Satellite Contract." *Washington Post,* December 17, 1965, C8.

———. "Early Bird Line Demand Grows." *Washington Post,* June 8, 1965, D8.

"Welch, Charyk Picked to Head Satellite Firm." *Washington Post,* March 1, 1963, A3.

Wentworth, Eric. "COMSAT Gyrations Scrutinized by Federal Agencies in Case a Sudden Price Plunge Sparks Public Furor." *Wall Street Journal,* December 22, 1964, 3.

Wilford, John Noble. "COMSAT Seeking Bigger Satellite." *New York Times,* December 30, 1965, 21.

Government Documents

Bullock, Gilbert D. "Applications Technology Satellite Program Summary." Revised April 1968. AA65-84. NASA Goddard Space Flight Center, NHO.

"Chronology of Communication Satellite Program." N.d. HAC.

Director of Defense Research and Engineering (DDRE). "Report: Management of Advent." April 3, 1960. JHR.

DoD (PA), news release. "NASA to Transfer Syncom Satellites to Defense Department." January 4, 1965.

———. "Syncom Communications Satellites Transfer Complete." #451-65. July 6, 1965.

Dryden, Hugh L. Press conference transcript. January 14, 1961, 21. NHO.

Hall, Harvey. "Early History and Background on Earth Satellites." ONR:405:HH:dr. November 29, 1957. NHO.

Harriman, Edward E. "Project Summary on Courier." November 18, 1960. NHO.

NASA. "Associate Administrator's Advanced Study Review: Earth Orbital Studies," A19–A20. June 25, 1964. NHO.

———. "Early Bird I Post-Launch Report No. 1." NASA MOR S-631-65-01. April 7, 1965. NHO.

———. "Early Bird I Post-Launch Report No. 2." NASA MOR S-631-65-01. April 16, 1965. NHO.

———. "HS-303 Early Bird Communications Satellite." NASA Mission Operation Report S-631-65-01. March 31, 1965. NHO.

———. "Program Management Report: Space Applications Programs." May 23, 1967. NASA HQ. NHO.

———. "Program Review: Communications, Meteorology, Future Applications." June 29, 1963. NHO.

NASA, news release. "Budget Briefing." #61-115. May 25, 1961. NHO.

———. "NASA Plans Follow-on Communications Satellite." #62-139. June 18, 1962. NHO.

———. "NASA Syncom Satellites Cross Paths 22,000 Miles above Pacific." #64-217. September 2, 1964. NHO.

———. "NASA to Transfer Syncom Satellites to Defense Department." #65-5. December 31, 1964. NHO.

———. "Statement by James E. Webb, Administrator." #61-112. 1961. May 25, 1961. NHO.

———. "Technical Background Briefing: Project Syncom." January 29, 1962. NHO.

———. "Technical Background Briefing: Project Syncom." January 29, 1963. NHO.

NASA, press release. September 1, 1960. NHO.

Newell, Homer, to J. E. Webb. "Intelsat II-A, Post Launch Report No. 1." NASA MOR S-631-65-02. November 2, 1966. NHO.

Thompson, George Raynor. "NASA's Role in the Development of Communications Satellite Technology." Unpublished NASA Historical Manuscript no. 8 (HHM-8). [November?] 1965. NHO.

U.S. Congress. House. *A Chronology of Missile and Astronautic Events [1915–1960]*. 87th Cong., 1st sess., 1961.

U.S. Congress. House. Committee on Aeronautical and Space Sciences. *Hearings: Commercial Communications Satellites*. September 18, 19, 21, 27, October 4, 1962. 87th Cong., 2d sess., 1962.

———. *Report: Communications Satellite Act of 1962*. 87th Cong., 2d sess., 1962.

U.S. Congress. House. Committee on Government Operations, Military Operations Subcommittee. *Report: Satellite Communications—Military-Civil Roles and Relationships*. 88th Cong., 2d sess., 1964.

U.S. Congress. House. Committee on Interstate and Foreign Commerce. *Hearings: Communications Satellites, Part 2*. 87th Cong., 1st sess., 1961.

U.S. Congress. House. Committee on Science and Astronautics. *Defense Space Interests*. 87th Cong., 1st sess., 1961.

———. *Hearings: Communications Satellites, Part I*. 87th Cong., 1st sess., 1961.

———. *Hearings: Communications Satellites, Part 2*. 87th Cong., 1st sess., 1961.

———. *Hearings: Project Advent—Military Communications Satellite Program*. 87th Cong., 2d sess., 1962.

———. *Hearings: Satellites for World Communication*. 86th Cong., 1st sess., 1959.

———. *Report: Commercial Applications of Space Communications Systems*. 87th Cong., 1st sess., 1961.

———. *Report: Satellites for World Communication*. 86th Cong., 1st sess., 1959.

U.S. Congress. House. Select Committee on Astronautics and Space Exploration. *The National Space Program*. 85th Cong., 2d sess., 1958.

———. *The Next Ten Years in Space, 1959–1969*. 86th Cong., 1st sess., 1959.

U.S. Congress. Office of Technology Assessment. *International Cooperation and Competition in Civilian Space Activities*. Washington, D.C.: GPO, 1985.

U.S. Congress. Senate. Committee on Aeronautical and Space Sciences. *Communications Satellites: Technical, Economic, and International Developments*. 87th Cong., 1st sess., 1962.

———. *Staff Report on Communications Satellites: Technical, Economic, and International Developments*. 87th Cong., 2d sess., 1962.

U.S. Congress. Senate. NASA Authorization Subcommittee of the Committee on Aeronautical and Space Sciences. *Hearings on NASA Authorization for Fiscal Year 1961*. 86th Cong., 2d sess., 1960.

U.S. Department of Commerce. *Historical Statistics of the United States: Colonial Times to 1970*. Washington, D.C.: GPO, 1975.

———. *1984 World's Submarine Telephone Cable Systems*. Washington, D.C.: GPO, 1984.

Weitzel, Ronald J. "The Origins of ATS." NASA Historical Note no. 83 (HHN-83). August 28, 1968. NHO.
The White House, press release. "Statement of the President on Communications Satellite Policy." July 24, 1961. NAII.

Unpublished Materials and Correspondence

Adler, F. P., to J. W. Ludwig, H. A. Rosen. "Visit of Mr. Jaffe." August 26, 1960. HAC.
AT&T/BTL. "Project Echo: Monthly Report No. 3, Contract NASW-110, December 1959." AT&T.
———. "Project Telstar: Preliminary Report, Telstar 1." July–September 1962. AT&T.
Best, G. L. "Memorandum: Satellite Communications." September 6, 1960. AT&T.
Best, G. L., to T. Keith Glennan. September 15, 1960. AT&T.
Betts, A. W., to J. H. Rubel. "Communications Satellite." June 13, 1960. JHR.
Brown, Harold, (John Rubel), to George P. Miller. July 10, 1962. JHR.
Cole, R. W., to D. D. Williams. "Summary of Orbital Data for Syncom 2 (A-26)." August 18, 1963. HAC.
"Commercial Communication Satellite, Preliminary Cost Estimates." September 14, 1960. HAC.
"Comsat Financial Status: 3 July 1960." July 13, 1960. HAC.
Comsat. "COMSAT at Fifteen." N.d. [late 1977]. Author's files.
———. "COMSAT, the First Ten Years: Report to the President and the Congress." July 31, 1973. Author's files.
———. "Report Pursuant to Section 404 (b) of the Communications Satellite Act of 1962 for the Period February 1, 1963, to December 31, 1963." January 31, 1964. Author's files.
Comsat, press release. May 26, 1965. January 25, 1965. NHO.
———. "Commercial Communications Satellite Engineering Design Proposals Requested of Industry; Corporation Regards Move as Major Step." December 22, 1963. Author's files.
———. "COMSAT Asks Manufacturers to Submit Proposals on Advanced Satellite for Global Communications." August 17, 1965. Author's files.
———. "COMSAT Board Elects James McCormack Chief Executive Officer and Board Chairman." October 15, 1965. Author's files.
———. "COMSAT Files Application for FCC Authorization for Proposed Satellites for Apollo Service." September 30, 1965. Author's files.
———. "COMSAT Gives Highlights of Corporation's Progress in an Interim Report to Shareholders." December 31, 1965. Author's files.
———. "COMSAT Shareholders to Elect Twelve Directors at May 11 Meeting." May 7, 1965. Author's files.
———. "Early Bird Fact Sheet I: Key Early Bird Project Officials." March 30, 1965. Author's files.
———. "Early Bird Fact Sheet III: Early Bird." March 30, 1965. Author's files.

———. "Note to Correspondents." December 10, 1963. Author's files.

———. "Welch Announces Wish to Retire as COMSAT Chief Executive Officer and Board Chairman." July 7, 1965. Author's files.

Corbo, R. L., to T. W. Oswald. "Revision of the Communication Satellite Structure and General Arrangement." January 29, 1960. HAC.

Corliss, William R. "History of the Goddard Networks." November 1, 1969. NHO.

Cunniffe, Peter. "Misreading History: Government Intervention in the Development of Commercial Communications Satellites." May 1991. Report #24, Program in Science and Technology for International Security, MIT, Cambridge.

Cutler, C . C., to J. R. Pierce. October 27, 1958. AT&T.

Darcey, R., to H. Goett. June 14, 1963. NHO.

Deputy Associate Administrator (Hilburn) to Associate Administrator (Seamans). "Possible Use of Early Bird in Support of the Apollo Network." July 19, 1965. NHO.

Dickieson, A. C. "TELSTAR: The Management Story." Unpublished monograph. Bell Telephone Laboratories, July 1970. AT&T.

Dingman, James E., to Leo Welch. December 6, 1963. NHO.

Doody, David F., to L. A. Hyland. "Commercial Communication Satellite (H. A. Rosen and D. D. Williams) Patentable Novelty." October 29, 1959. HAC.

Doody, David F., to Noel Hammond. "Communication Satellite." December 1, 1959. HAC.

Douglas, James H. (acting Secretary of Defense), to Secretary of the Army and Secretary of the Air Force. "Program Management for ADVENT." September 15, 1960. JHR.

Dryden, H. L. Press conference transcript. January 14, 1961, 21. NHO.

Dryden, H. L., to J. V. Charyk. June 24, 1963. NHO.

Field, L. M., to L. A. Hyland. "Possible Use of Syncom as a Navigational System—Microwave Loran." August 9, 1963. HAC.

Ford Aerospace and Communications Corporation. "Intelsat V Spacecraft System Summary." September 21, 1977. Author's files.

Frutkin, Arnold W., to Eugene M. Emme. December 30, 1965. NHO.

Garbarini, R., to H. Goett. September 24, 1963. NHO.

Glennan, T. Keith, to G. L. Best. September 28, 1960. AT&T.

Glennan, T. Keith, to James H. Douglas. August 18, 1960. NHO.

Glennan, T. Keith, to J. Johnson. January 5, 1959. NHO.

GM-Hughes Electronics. "History and Accomplishments of the Hughes Aircraft Company." N.d. HAC.

Green, E. I., to T. Keith Glennan. October 20, 1960. AT&T.

Hazelrigg, George A., Jr., and Eli B. Roth. "Windows for Innovation: A Story of Two Large-Scale Technologies." Econ, Inc., Report to NSF. December 1, 1982. Author's files.

Hirsh, Richard. "Memorandum for Dr. Welsh: Highlights of Meeting of June 27, 1961, Concerning Communications Satellites." June 28, 1961. NAII.

———. "Memorandum for Dr. Welsh: Highlights of Meeting of June 28, 1961, Concerning Communications Satellites." June 29, 1961. NAII.

Hughes Aircraft Company. "Advantages of Stationary Satellite Systems." N.d. HAC.

———. "Comparison of the Hughes Synchronous Repeater with the BTL Low Altitude Repeater." N.d. HAC.

———. "Synchronous Communication Satellite, Proposed NASA Experimental Program, Exhibit B: Technical Discussion." ED 655R, 60H-6702/5874-005. June 1960. HAC.

———. "SYNCOM" (brochure). 3/62/2M. NHO.

Humphrey, Kefauver, Morse, et al. to John F. Kennedy, August 24, 1961. Reprinted in U.S. Congress, Senate, Committee on Foreign Relations, *Hearings: Communications Satellite Act of 1962,* 87th Cong., 2d sess., 1962, 51–54.

Hyland, L. A., to A. V. Haeff, C. G. Murphy. "Communications Satellite." October 26, 1959. HAC.

Hyland, L. A., to Robert Gilruth. October 21, 1963. HAC.

Iams, John D., to Arnold Frutkin. December 9, 1965. NHO.

Jaffe, Leonard, to Eugene M. Emme. February 24, 1966. NHO.

Jakes, W. C. "Visit to Washington on March 31, 1959." April 7, 1959. AT&T.

Jerrems, A. S., to F. R. Carver. September 17, 1959. HAC.

Jerrems, A. S., to H. A. Rosen. "Syncom Publicity." September 21, 1964. HAC.

Johnson, Roy W., to T. Keith Glennan. December 17, 1958. NHO.

Kappel, Fred R., to James E. Webb. April 5, May 8, 1961. Copies in *Exploring the Unknown: Selected Documents in the History of the U.S. Civil Space Program,* edited by John M. Logsdon et al., vol. 3., *Using Space.* Washington, D.C.: NASA, 1998.

Kecskemeti, Paul. "The Satellite Rocket Vehicle: Political and Psychological Problems." RAND RM-567. October 4, 1950. RAND, Santa Monica, Calif.

Kelleher, John J., memorandum to Leonard Jaffe. "FCC Action on Early Bird." January 24, 1964. NHO.

Kennedy, John F., draft of letter to LBJ. [June 1961.] In Welsh files. NAII.

Ludwig, J. W., to A. E. Puckett. "Army RFP on Advent System, Assignment of HAC Responsibility." May 19, 1960. HAC.

Ludwig, J. W., to C. G. Murphy/E. A. Puckett [*sic*]. "Communication Satellite—Presentation to Dr. E. G. Wittimg, Deputy Director, R and D, U.S. Army." May 2, 1960. HAC.

Ludwig, J. W., and J. H. Striebel to R. K. Roney. "Commercial Communications Satellite—Logistics." December 22, 1959. HAC.

Lutz, S. G., to A. V. Haeff. "Commercial Satellite Communication Project: Preliminary Report of Study Task Force." October 22, 1959. HAC.

———. "Economic Aspects of Satellite Communication." October 13, 1959. HAC.

———. "Evaluation of H. A. Rosen's Commercial Satellite Communication Proposal." October 1, 1959. HAC.

———. "Satellite Data." October 29, 1959. HAC.

Lutz, S. G., to file. "AFCEA Papers by B.T.L. on Satellite Communication." June 3, 1960. HAC.

———. "Conference with Leon Jaffee [*sic*], NASA, Re Satellite Communication, Frequency Sharing." May 31, 1960. HAC.

Mathews, Wayne C., to J. E. Webb. "Intelsat-II-A Post launch Report No. 2." MOR S-631-65-02. November 7, 1966. NHO.

McKay, K. G., to Ellis Rubinstein. November 12, 1976. AT&T.

Meckling, William, and Siegfried Reiger. "Communications Satellites: An Introductory Survey of Technology and Economic Promise." RAND Corporation, Report RM-2709-NASA. September 15, 1960. RAND, Santa Monica, Calif.

Minutes: Administrator's Staff Meeting. January 18, 26, February 2, May 25, June 1, 8, 15, 22, 29, 1961. NHO.

Morse, Edgar W. "Preliminary History of the Origins of Project Syncom." NASA Historical Note no. 44 (HHN-44). September 1, 1964. NHO.

[Murphy, C. G.?]. "Policy Statement for Exploitation of HAC Communications Satellites." [Early 1962?]. HAC.

Murphy, C. G., to H. A. Rosen. "Visit of Dr. S. Reiger and Other RAND Corporation Personnel." September 21, 1960. HAC.

Murphy, C. G., to L. A. Hyland. "Synchronous Altitude Communication Satellite System." September 16, 1963. HAC.

Naugle, John, memorandum to Administrator. "Recommendation for Executive Performance Award." July 28, 1975. NHO.

Newell, Homer, to J. E. Webb. "Intelsat II-A, Post Launch Report No. 1." MOR S-631-65-02. November 2, 1966. NHO.

Nunn, Robert G. Memorandum. October 28, 1960. NHO.

———. "Memorandum for Record." May 5, 1961. NHO.

———. "Memorandum for the Associate Administrator." May 16, 1961. NHO.

———. "Memorandum for the Record." December 23, 1960. NHO.

———. Memorandum to Associate Administrator [Robert Seamans]. May 16, 1961. NHO.

Nunn, Robert G., to Eugene M. Emme. November 29, 1965. NHO.

O'Neill, E. F., to J. R. Pierce. August 17, 18, 1977. AT&T.

Ostrander, Don, to Dr. Seamans. "Reflections on the American Posture in Space." April 21, 1961. NHO.

Penzias, Arno A., to J. R. Pierce. July 8, 1976. July 8, 1977. AT&T.

Pierce, J. R. (?) "BTL Tracking Proposal." April 2, 1959. AT&T.

Pierce, J. R., to C. C. Cutler. October 17, 1958. AT&T.

Pierce, J. R., to E. Rubinstein. November 2, December 17, 1976. AT&T.

Pierce, J. R., to Eugene M. Emme. January 7, 1966. NHO.

Pierce, J. R., to J. A. Morton. April 14, 1959. AT&T.

Pierce, J. R., to J. W. McRae. January 7, 1959. AT&T.

Pierce, J. R., to K. G. McKay. November 23, December 17, 1976. AT&T.

Pierce, J. R., R. Kompfner, and C. C. Cutler. "Memorandum for the Record: Research toward Satellite Communication, Case 38543." January 6, 9, 1959. AT&T.

Puckett, Allen E., to D. F. Doody. "Release of Hughes' Rights in Inventions Disclosed Relating to Communications Satellite." March 7, 1960. HAC.

Puckett, Allen E., to L. A. Hyland. "Communication Satellite." March 21, 1960. HAC.

Puckett, Allen E., to [J. W.] Ludwig, [H. A.] Rosen. "Visit from ITT Laboratories Regarding Communication Satellite." November 18, 1960. HAC.
Puckett, Allen E., to Richard S. Morse. April 8, 1960. HAC.
Puckett, Allen E., to Those Listed. "Communication Satellite Presentation." July 26, 1960. HAC.
Radius, Walter A., to ADA/Willis Shapley. December 10, 1965. NHO.
RAND. "A Comparison of Long Range Surface-to-Surface Rocket and Ramjet Missiles." R-174. 1950. RAND, Santa Monica, Calif.
———. "Preliminary Design of an Experimental World-Circling Spaceship." SM-11827. May 12, 1946. RAND, Santa Monica, Calif.
———. "Satellite to Surface Communication: Equatorial Orbit." RM-603. July 1949. RAND, Santa Monica, Calif.
Reiger, S. H., R. T. Nichols, L. B. Early, and E. Dews. "Communications Satellites: Technology, Economics, and System Choices." RM-3487-RC. February 1963. RAND, Santa Monica, Calif.
Richardson, John S., to C. G. Murphy. "Communications Satellite." August 12, 1960. HAC.
Roney, Robert K., to A. E. Puckett. "Communications Satellite Review Analysis." January 27, 1960. HAC.
Rosen, H. A., and D. D. Williams. "Commercial Communication Satellite." October 1959. Airborne Systems Laboratories, Hughes Aircraft Company. HAC.
———. "Commercial Communication Satellite." RDL/B-1. January 1960. Engineering Division, Hughes Aircraft Company. HAC.
Rosen, H. A., and Tom Hudspeth. Interview. California Museum of Science and Technology. Spring 1992. HAC.
Rosen, H. A., to E. G. Witting. October 25, 1960. HAC.
Rosen, H. A., to L. A. Hyland. "Syncom Personnel." September 23, 1964. HAC.
Rubel, John H., to A. S. Jerrems. September 22, 1960. HAC.
Rubel, John H., to T. Keith Glennan. September 27, 1960. JHR.
Rubinstein, Ellis, to J. R. Pierce. December 9, 1976. AT&T.
Schwebs, Dinter H. (IDA/ARPA). "Memorandum for the Record: Reorientation of NOTUS Program." December 8, 1959. NHO.
Seamans, Robert C. NASA Exit Interview. May 8, May 30, and June 3, 1968. NHO.
Seamans, Robert C., Jr., to Dr. Joseph V. Charyk, November 23, December 30, 1965. NHO.
Seamans, Robert C., to [J. E.] Webb and [H. L.] Dryden. "Status of Planning for an Accelerated NASA Program." May 12, 1961. NHO.
Silverstein, Abe, to Assistant Directors et al. "Fiscal Year 1963 Preliminary Budget Estimates: Additional Information Concerning." March 1, 1961. NHO.
Striebel, J. H., to A. V. Haeff. "Market Study of a World Wide Communication System for Commercial Use." October 22, 1959. HAC.
Thompson, George Raynor. "NASA's Role in the Development of Communications Satellite Technology." (HHM-8). [November?] 1965. NHO.

Vieth, F. D., to Lt. Col. L. B. Brownfield. May 27, 1960. HAC.
Vogel, Lou, to Eugene M. Emme. December 13, 1965. NHO.
Warren, R. E. Memorandum. October 14, 1960. NHO.
———. "Syncom I Progress Report No. 4." March 4, 1963. NHO.
Warren, R. E., to R. Garbarini. September 10, 1963. NHO.
Webb, James E. "Administrator's Presentation to the President." March 21, 1961. NHO.
———. "Memorandum for Dr. Dryden." June 12, 1961. NHO.
Webb, James E., to Fred R. Kappel. April 8, 1961. Copy in *Exploring the Unknown: Selected Documents in the History of the U.S. Civil Space Program,* edited by John M. Logsdon et al., vol. 3., *Using Space.* Washington, D.C.: NASA, 1998.
Webb, James E., to John A. Johnson. April 28, 1961. NHO.
Webb, James E., to Overton Brooks. May 2, 1961. NHO.
Webb, James E., to Robert S. McNamara. June 1, 1961. NHO.
Webb, James E., to W. Shapley. August 11, 12, 1965. NHO.
Webb, James E., to "The Director," Bureau of the Budget. March 13, 1961. NHO.
Weber, Fred. "INTELSAT Satellite History." COMSAT Memorandum. March 31, 1980. NHO.
Weber, W. J. "Memorandum for the Record: Summary of HS-303, 'Early Bird' Communications Satellite Program." January 15, 1965. NHO.
Welber, I. "Memorandum for File." July 10, 1961. AT&T.
Welsh, Edward C. "Memorandum for the Vice President." April 28, June 5, 1961. NAII.
Williams, D. D. "Dynamic Analysis and Design of the Synchronous Communication Satellite." TM-649. May 1960. HAC.
Williams, D. D., and P. Wong to F. R. Carver. "Orbit Determination for Satellite Surveillance System." August 18, 1960. HAC.
Williams, D. D., to D. F. Doody. "Discussions with Dr. Homer J. Stewart, NASA." November 23, 1959. HAC.
Williams, D. D., to Noel B. Hammond. "Information Related to PD-4286, Velocity Control, and Orientation of a Spin-Stabilized Body." July 27, 1964. HAC.

Index

Page numbers in *italics* refer to illustrations